T0324928

SPECTRAL THEORY AND ITS APPLICATIONS

Bernard Helffer's graduate-level introduction to the basic tools of spectral analysis is illustrated by numerous examples from the theory of Schrödinger operators and various branches of physics, including statistical mechanics, superconductivity, fluid mechanics, and kinetic theory. The later chapters also introduce the theory of non-self-adjoint operators, with an emphasis on the role of pseudospectra.

The author's focus on applications, along with exercises and examples, enables readers to connect theory with practice so that they will develop a good understanding of how the abstract spectral theory can be applied. The final chapter provides various problems that have been the subject of active research in recent years and will challenge the reader's understanding of the material covered.

Bernard Helffer is a Professor in the Department of Mathematics at Université Paris-Sud. He has published more than 200 papers in mathematics and mathematical physics and authored five books. In 2011 he was awarded the Prix de l'État by the French Academy of Sciences.

Spectral Theory and its Applications

BERNARD HELFFER
Université Paris-Sud

CAMBRIDGE
UNIVERSITY PRESS

CAMBRIDGE
UNIVERSITY PRESS

University Printing House, Cambridge CB2 8BS, United Kingdom

Cambridge University Press is part of the University of Cambridge.

It furthers the University's mission by disseminating knowledge in the pursuit of education, learning and research at the highest international levels of excellence.

www.cambridge.org
Information on this title: www.cambridge.org/9781107032309

First published 2013

A catalogue record for this publication is available from the British Library

Library of Congress Cataloguing in Publication data
Helffer, Bernard.
Spectral theory and its applications / Bernard Helffer.
pages   cm. – (Cambridge studies in advanced mathematics ; 139)
ISBN 978-1-107-03230-9 (hardback)
1. Spectral theory (Mathematics)   I. Title.
QC20.7.S64H45   2013
515'.7222–dc23
2012028839

ISBN 978-1-107-03230-9 Hardback

Contents

Contents

1

Introduction

1.1 Spectral theory in action

In this book, we present the basic tools of spectral analysis and illustrate the theory by presenting many examples from the theory of Schrödinger operators and from various branches of physics, including statistical mechanics, super-conductivity, fluid mechanics, and kinetic theory. Hence we shall alternately present parts of the theory and use applications in those fields as examples. In the final chapters, we also give an introduction to the theory of non-self-adjoint operators with an emphasis on the role of pseudospectra. Throughout the book, the reader is assumed to have some elementary knowledge of Hilbertian and functional analysis and, for many examples and exercises, to have had some practice in distribution theory and Sobolev spaces. This introduction is intended to be a rather informal walk through some questions in spectral theory. We shall answer these questions mainly "by hand" using examples, with the aim of showing the need for a general theory to explain the results. Only in Chapter 2 will we start to give precise definitions and statements.

Our starting point is the theory of Hermitian matrices, that is, the theory of matrices satisfying $A^\star = A$, where $A^\star$ is the adjoint matrix of A. When we are looking for eigenvectors and corresponding eigenvalues of A, that is, for pairs (u, λ) with $u \in \mathbb{C}^k$, $u \neq 0$, and $\lambda \in \mathbb{C}$ such that $Au = \lambda u$, we know that the eigenvalues will be real and that one can find an orthonormal basis of eigenvectors associated with those eigenvalues. In this case, we can speak of eigenpairs.

In order to extend this theory to the case of spaces with infinite dimension (that is, where the space $\mathbb{C}^m$ is replaced by a general Hilbert space $\mathcal{H}$), we might attempt to develop a theory of compact self-adjoint operators. But it would be a major task to cover all the interesting cases that arise in quantum mechanics. So, although our aim is to present a general theory, it is perhaps good to start by looking at specific operators and asking naive questions about the existence of eigenpairs (u, λ) with u in some suitable domain, $u \neq 0$ and

1

$\lambda \in \mathbb{C}$, such that $Au = \lambda u$. We shall discover in particular that the answers to these questions may depend strongly on the choice of the domain and on the precise definition of the operator.

1.2 The free Laplacian

In this spirit, let us start with the free Laplacian in $\mathbb{R}^m$. We denote by $L^2(\mathbb{R}^m)$ the space of (or class of) measurable functions on $\mathbb{R}^m$ that are square integrable with respect to the Lebesgue measure dx (for $dx_1, \ldots, dx_m$). The Laplacian

$$-\Delta = -\sum_{j=1}^m \frac{\partial^2}{\partial x_j^2}$$

has no eigenfunctions in $L^2(\mathbb{R}^m)$, i.e., there is not a pair (u, λ) with $\lambda \in \mathbb{C}$ and $u \neq 0$ in L^2 such that $-\Delta u = \lambda u$ in the sense of distributions.[1] But it has, for any $\lambda \in \mathbb{R}^+$, an eigenfunction in $\mathcal{S}'(\mathbb{R}^m)$ (the space of tempered distributions) (actually, in $L^\infty(\mathbb{R}^m)$) and, for any $\lambda \in \mathbb{C}$, an eigenfunction in $\mathcal{D}'(\mathbb{R}^m)$ (the space of distributions). So what is the right way to extend the theory of Hermitian matrices on $\mathbb{C}^k$?

On the other hand, it is easy to produce approximate eigenfunctions of the form $u_n(x) = n^{-m/2} e^{ix \cdot \xi} \chi(x/n)$, where χ is a compactly supported C^∞ function with an L^2-norm equal to 1 and $\xi \in \mathbb{R}^m$. By "approximate" we mean that if $\lambda = |\xi|^2$ and $A = -\Delta$, the norm in L^2 of $(A - \lambda)u_n$ tends to 0 as $n \to +\infty$.

1.3 The harmonic oscillator

As we shall see, the operator of the harmonic oscillator (referred to simply as the "harmonic oscillator" from now on),

$$H = -\frac{d^2}{dx^2} + x^2,$$

plays a central role in the theory of quantum mechanics. When we look for eigenfunctions in $\mathcal{S}(\mathbb{R})$ (the Schwartz space of C^∞ rapidly decreasing functions at ∞, together with all derivatives), we can show that there is a sequence of eigenvalues λ_n $(n \in \mathbb{N}^*)$,

$$\lambda_n = (2n - 1).$$

---

[1] This means in this case that $-\int u(x)(\Delta\phi)\,dx = \lambda \int u(x)\phi(x)\,dx$, for any function ϕ in $C_0^\infty(\mathbb{R}^m)$. Some other authors use the notion of a weak solution.

In particular, the fundamental level (in other words, the lowest eigenvalue) is $\lambda_1 = 1$ and the splitting between the first two eigenvalues is 2.

The first (normalized) eigenfunction is given by

$$\phi_1(x) = \pi^{-1/2} \exp -\frac{x^2}{2}, \tag{1.3.1}$$

and the other eigenfunctions are obtained by applying the so-called[2] creation operator

$$L^+ = -\frac{d}{dx} + x. \tag{1.3.2}$$

We observe that

$$H = L^+ \cdot L^- + 1, \tag{1.3.3}$$

where

$$L^- = \frac{d}{dx} + x, \tag{1.3.4}$$

and L^- has the property

$$L^- \phi_1 = 0. \tag{1.3.5}$$

Note that if $u \in L^2$ is a distributional solution of $L^+ u = 0$, then $u = 0$. Also, if $u \in L^2$ is a distributional solution of $L^- u = 0$, then $u = \mu \phi_1$ for some $\mu \in \mathbb{R}$.

The nth eigenfunction is then given by

$$\phi_n = 2^{-(n-1)/2} ((n-1)!)^{-1/2} (L^+)^{n-1} \phi_1. \tag{1.3.6}$$

This can be shown by recursion using the identity

$$L^+ (H + 2) = HL^+. \tag{1.3.7}$$

It is easy to see that

$$\phi_n(x) = P_n(x) \exp -\frac{x^2}{2}, \tag{1.3.8}$$

where $P_n(x)$ is a polynomial of order $n - 1$. It can also be shown that the ϕ_n are mutually orthogonal. The proof of this point is identical to the

---

[2] In quantum mechanics.

finite-dimensional case, if we observe the following identity (expressing the fact that H is symmetric):

$$\langle Hu, v \rangle_{L^2} = \langle u, Hv \rangle_{L^2}, \forall u \in \mathcal{S}(\mathbb{R}), \forall v \in \mathcal{S}(\mathbb{R}), \qquad (1.3.9)$$

which is obtained by an integration by parts.

We also observe that, by recursion, $||\phi_n|| = 1$. It is then a standard exercise to show that the family $(\phi_n)_{n \in \mathbb{N}}$ is total in $L^2(\mathbb{R})$ (i.e., the vector space generated by finite linear combinations of elements of the family is dense in L^2). A direct way is to analyze, for any $g \in L^2$, the function $\mathbb{R} \ni \xi \mapsto F_g(\xi) = \int_{\mathbb{R}} \exp -ix\xi \, g(x)\phi_1(x) \, dx$, and to observe that, owing to the Gaussian decay of ϕ_1, this is a real analytic function on $\mathbb{R}$. Moreover, if g is orthogonal to all of the ϕ_n, then $F_g^{(k)}(0) = 0$ for any k. This implies $F_g(\xi) = 0, \forall \xi \in \mathbb{R}$. But $F_g(\xi)$ is the Fourier transform of $g\phi_1$, and hence $g = 0$. Hence we have obtained an orthonormal Hilbertian basis of $L^2(\mathbb{R})$, which in some sense permits us to diagonalize the operator H.

Another way to understand this completeness is to show that if we start with an eigenfunction u in $\mathcal{S}'(\mathbb{R})$ associated with $\lambda \in \mathbb{R}$ that is a solution (in the sense of distributions) of

$$Hu = \lambda u,$$

then there exist $k \in \mathbb{N}$ and $c_k \neq 0$ such that $(L^-)^k u = c_k \phi_1$ and that the corresponding λ is equal to $(2k + 1)$. For the proof of this, we have to assume that any eigenfunction is in $\mathcal{S}(\mathbb{R})$ (this can be proven independently of any explicit knowledge of the eigenfunctions) and use the identity

$$L^-(H - 2) = HL^- \qquad (1.3.10)$$

and the inequality

$$\langle Hu, u \rangle \geq 0, \forall u \in \mathcal{S}(\mathbb{R}). \qquad (1.3.11)$$

This last property is called the "nonnegativity" of the operator.

Actually, it can be shown in various ways that

$$\langle Hu, u \rangle \geq ||u||^2, \forall u \in \mathcal{S}(\mathbb{R}). \qquad (1.3.12)$$

One way is to first establish the Heisenberg uncertainty principle,

$$||u||^2_{L^2(\mathbb{R})} \leq 2||xu||_{L^2}||u'||_{L^2}, \forall u \in \mathcal{S}(\mathbb{R}). \qquad (1.3.13)$$

Before we describe the trick behind the proof, however, let us give a more "physical" version.

If u is normalized by $||u||_{L^2(\mathbb{R})} = 1$, the measure $|u|^2 \, dx$ is a probability measure. One can then define the mean value of the position by

$$\langle x \rangle = \int x |u|^2 \, dx$$

and the variance by

$$\sigma_x = \langle (x - \langle x \rangle)^2 \rangle.$$

Similarly, we can consider

$$\langle D_x \rangle := \int (D_x u) \cdot \bar{u}(x) \, dx$$

(with $D_x = -i d/dx$) and

$$\sigma_{D_x} := ||(D_x - \langle D_x \rangle) u||^2.$$

Then (1.3.13) can be extended in the form

$$\sigma_x \cdot \sigma_{D_x} \geq \frac{1}{4}.$$

The trick is to observe the identity

$$1 = \frac{d}{dx} \cdot x - x \cdot \frac{d}{dx}. \tag{1.3.14}$$

We then write, for $u \in \mathcal{S}(\mathbb{R})$,

$$u(x) \, \bar{u}(x) = \left(\left(\frac{d}{dx} \cdot x - x \cdot \frac{d}{dx} \right) u(x) \right) \bar{u}(x),$$

and then integrate over $\mathbb{R}$:

$$\int_{\mathbb{R}} |u(x)|^2 \, dx = \int (xu)' \, \bar{u}(x) \, dx - \int xu'(x) \, \bar{u}(x) \, dx.$$

After an integration by parts, we obtain

$$\int_{\mathbb{R}} |u(x)|^2 \, dx = - \int xu'(x) \bar{u}(x) \, dx - \int xu(x) \bar{u}'(x) \, dx.$$

(1.3.13) is then a consequence of the Cauchy–Schwarz inequality.

The inequality (1.3.12) is simply a consequence of the identity

$$\langle Hu, \, u \rangle = ||u'||^2 + ||xu||^2, \tag{1.3.15}$$

which can be proved by an integration by parts, and of the application in (1.3.13) of the Cauchy–Schwarz inequality. Another way is to directly observe the identity

$$\langle Hu, \, u \rangle = ||L^- u||^2 + ||u||^2, \; \forall u \in \mathcal{S}(\mathbb{R}). \tag{1.3.16}$$

1.4 The problem of the boundary

We consider mainly ordinary differential operators of first or second order on an interval $]0, 1[$ and look at various questions that can be asked naively about the existence of eigenfunctions for the problem in $L^2(]0, 1[)$.

1.4.1 Preliminary discussion

We look first at pairs $(u, \lambda) \in H^1(]0, 1[) \times \mathbb{C} \, (u \neq 0)$ such that

$$-\frac{du}{dx} = \lambda u, \; u(0) = 0,$$

where $H^1(]0, 1[)$ is the Sobolev space

$$H^1(]0, 1[) = \{u \in L^2(]0, 1[) \, | \, u' \in L^2(]0, 1[)\}.$$

Here we recall that $H^1(]0, 1[)$ is included in $C^0([0, 1])$, by the Sobolev injection theorem. It can be seen immediately that no such pairs exist. We shall come back to this example later when we analyze non-self-adjoint problems.

We now look at pairs $(u, \lambda) \in H^2(]0, 1[) \times \mathbb{C} \, (u \neq 0)$ such that

$$-\frac{d^2 u}{dx^2} = \lambda u.$$

For any λ, we can find two linearly independent solutions.

1.4.2 The periodic problem

We consider pairs $(u, \lambda) \in H^{2, \mathrm{per}}(]0, 1[) \times \mathbb{C} \, (u \neq 0)$ such that

$$-\frac{d^2 u}{dx^2} = \lambda u.$$

Here,

$$H^{2,\mathrm{per}}(]0, 1[) = \{u \in H^2(]0, 1[), \ u(0) = u(1) \text{ and } u'(0) = u'(1)\},$$

where $H^2(]0, 1[)$ is the Sobolev space

$$H^2(]0, 1[) = \{u \in H^1(]0, 1[) \mid u' \in H^1(]0, 1[)\}.$$

Here we recall that $H^2(]0, 1[)$ is included in $C^1([0, 1])$, by the Sobolev injection theorem. It is an easy exercise to show that the pairs are described by two families:

- $\lambda = 4\pi^2 n^2$, $u_n = \mu \cos 2\pi nx$, for $n \in \mathbb{N}$, $\mu \in \mathbb{R} \setminus \{0\}$,
- $\lambda = 4\pi^2 n^2$, $v_n = \mu \sin 2\pi nx$, for $n \in \mathbb{N}^*$, $\mu \in \mathbb{R} \setminus \{0\}$.

We observe that $\lambda = 0$ is the lowest eigenvalue and that its multiplicity is one. This means that the corresponding eigenspace is of dimension one (the other eigenspaces are of dimension 2). Moreover, an eigenfunction in this subspace never vanishes in $]0, 1[$. This is quite evident, because $u_0 = \mu \neq 0$. One can also obtain an orthonormal basis of eigenfunctions in $L^2(]0, 1[)$ by normalizing the family $(\cos 2\pi nx \ (n \in \mathbb{N}), \sin 2\pi nx \ (n \in \mathbb{N}^*))$ or the family $\exp 2\pi inx \ (n \in \mathbb{Z})$.

We are merely recovering the L^2-theory of Fourier series here.

1.4.3 The Dirichlet problem

Here we consider spectral pairs $(u, \lambda) \in H^{2,D}(]0, 1[) \times \mathbb{C} \ (u \neq 0)$ such that $-d^2u/dx^2 = \lambda u$, with $H^{2,D}(]0, 1[) = \{u \in H^2(]0, 1[), \ u(0) = u(1) = 0\}$. It is again an easy exercise to show that these pairs are described by

$$\lambda = \pi^2 n^2, \ v_n = \mu \sin \pi nx, \ \text{for} \ n \in \mathbb{N}^*, \ \mu \in \mathbb{R} \setminus \{0\}.$$

We observe that $\lambda = \pi^2$ is the lowest eigenvalue, that its multiplicity is one (here, all the eigenspaces are one dimensional), and that an eigenfunction in this subspace does not vanish in $]0, 1[$.

1.4.4 The Neumann problem

Here we consider eigenpairs $(u, \lambda) \in H^{2,N}(]0, 1[) \times \mathbb{C} \ (u \neq 0)$, such that

$$-d^2u/dx^2 = \lambda u,$$

where

$$H^{2,N}(]0, 1[) = \{u \in H^2(]0, 1[), \ u'(0) = u'(1) = 0\}.$$

It is again an easy exercise to show that these pairs are described by

$$\lambda = \pi^2 n^2, \ v_n = \mu \cos \pi n x, \ \text{for} \ n \in \mathbb{N}, \ \mu \in \mathbb{R} \setminus \{0\}.$$

We observe that $\lambda = 0$ is the lowest eigenvalue, that its multiplicity is one (here, again, all the eigenspaces are one dimensional), and that the corresponding eigenspace is of dimension one and that an eigenfunction in this subspace does not vanish in $]0, 1[$.

1.5 Aim and organization of the book

By looking at some rather simple operators, we have demonstrated various problems that occur when one tries to extend the notion of an eigenpair of a matrix. Many of the examples involving second-order ordinary differential operators can be treated by the so-called Sturm–Liouville theory. Our goal in this book is to develop a theory that is not limited to 1D problems and not based on explicit computations.

The book is organized into 16 chapters, which are mainly of two types: the first type presents an introduction to a theory (we sometimes choose to skip some of the proofs presented in standard, more complete textbooks), and the second type presents applications, without being afraid to give explicit computations. Hence we hope that the reader will always see how the theory can be used.

Chapter 2 is an introduction to the theory of unbounded operators. We assume here that the reader is familiar with basic Hilbertian and Banach theory, together with the theory of linear bounded operators on Hilbert and Banach spaces.

Chapter 3 presents the Lax–Milgram theorem, which is simply a useful variant of Riesz's theorem characterizing the dual of a Hilbert space.

Chapter 4 introduces the notion of semibounded operators (most of the time semibounded from below) and treats many examples, mainly from quantum mechanics. On the way, the reader will encounter inequalities that belong to the common background of the analyst, such as Hardy's inequality and Kato's inequality, which are of interest in their own right.

Chapter 5 recalls some rather basic material on compact operators and emphasizes examples. We also recall some tools from functional analysis that permit one to recognize if an operator is compact (precompactness criteria).

Chapter 6 also presents some basic material, about the spectral theory of bounded operators. This also provides us with an occasion to visit (sometimes only briefly) various aspects of functional analysis, such as Fredholm theory, index theory, subclasses of the space of compact operators, and the Krein–Rutman theorem.

Chapter 7 describes applications. The first of these is from statistical mechanics (continuous models). The other applications describe cases where the operators involved are unbounded but where one can come back, by considering the inverse, to the spectral theory of compact operators recalled in Chapters 5 and 6.

Chapter 8 returns to the more general spectral theory of unbounded operators, and is mainly standard and devoted to the presentation of an almost complete proof of the spectral theorem. We focus on elementary consequences of this theorem for functional calculus and for the determination of approximate eigenvalues.

Chapter 9, in some sense, answers some of the questions asked in Chapter 1. We give some rather explicit criteria for determining whether an operator is self-adjoint or whether it can be naturally extended to a self-adjoint operator. Again, we discuss some examples (mainly Schrödinger operators) thoroughly.

There are several different ways to distinguish between different subsets of a spectrum. We have chosen in Chapter 10 to present the decomposition of a spectrum into two parts: the essential spectrum and the discrete spectrum. As an application, we consider the case of the Schrödinger operator with a constant magnetic field.

In many situations, the problems considered depend on various parameters, and a comparison of the spectra of different problems could be difficult without a variational characterization of the eigenvalues, which is the subject of Chapter 11. This point of view is useful for finite-dimensional matrices.

Chapter 12 presents a short walk through the theory of fluid mechanics, and this permits us to see "spectral theory in action" with a rather explicit computation.

Chapter 13 is devoted to some ideas that appear to be important when we are no longer in the self-adjoint case. We give some basic properties of some specific families of neighborhoods of the spectrum computed by analyzing the growth of the resolvent. After recalling some elements of the theory of semigroups and their generators, we present some recent improvements on the Gearhart–Prüss theorem.

Since most of the books devoted to the material presented in Chapter 13 are written in a rather abstract way, we have chosen in Chapters 14 and 15 to discuss how the theorems can be applied in interesting cases. Here, "interesting" may mean either that one can compute everything in great detail or that an apparently simple model plays an important role in the explanation of a physical phenomenon, for example in the theory of superconductivity, in fluid mechanics, or in kinetic theory.

Although many examples are given in the body of and at the end of each chapter, in the last chapter we present various problems that should be solvable after the reader has understood, say, the first ten chapters.

Instead of giving references in the body of the chapters, we give references together with remarks and comments at the end of each chapter, in a section entitled "Notes."

Acknowledgments   Some of the material for this book was prepared together with colleagues for specific lectures or courses. I would like particularly to thank S. Fournais in this respect. Some other material has come rather directly from collaborations and discussions with Ph.D. students and colleagues: I would like in particular to thank Y. Almog, W. Bordeaux-Montrieux, T. Gallay, R. Henry, J. Martinet, F. Nier, X. Pan, and J. Sjöstrand for the part related to non-self-adjoint problems. Discussions (sometimes a long time ago) with B. Davies, A. Laptev, O. Lafitte, P. Lévy-Bruhl, and D. Robert were also very useful. Finally, I would like to thank R. Henry, F. Hérau, and C. Lena for their critical reading of chapters of the book.

2

Unbounded operators, adjoints, and self-adjoint operators

The development of spectral theory is strongly related to quantum mechanics, and the main operators that immediately appear in the theory are the operators of multiplication by x (in, say, $L^2(\mathbb{R})$), the operator of differentiation d/dx, and the harmonic oscillator $-d^2/dx^2 + x^2$. These operators are unbounded and, in fact, not defined for any element of $L^2(\mathbb{R})$. Of course, we can start by simply restricting the operator by introducing a smaller domain in the Hilbert space $L^2(\mathbb{R})$ of definition, but what is the right notion of continuity for the operator? How do we choose the "maximal" domain of definition? This is what we shall start to explain in this chapter.

2.1 Unbounded operators

We consider a Hilbert space $\mathcal{H}$. We assume that the reader has some basic knowledge of Hilbertian theory. The scalar product will be denoted by $\langle u, v \rangle_{\mathcal{H}}$ or, more simply, by $\langle u, v \rangle$ when no confusion is possible. We adopt the convention that the scalar product is antilinear with respect to the second argument.

We recall that a Hilbert space has a natural associated norm given by $\mathcal{H} \ni u \mapsto \sqrt{\langle u, u \rangle}$ and that with this structure it becomes a normed space, with the additional property that it is complete, i.e., in other words, it is a Banach space.

Our basic Hilbert spaces are the spaces $\ell^2(\mathbb{Z})$, $L^2(\mathbb{R}^m)$ or, more generally, the Sobolev spaces (which will be defined in (2.1.4) and (2.1.8)). They are all separable (that is, they have a countable dense subset), and hence they have an orthonormal basis (which can be labeled by $\mathbb{N}$), i.e., a family $(e_n)_{n \in \mathbb{N}}$ such that $\langle e_i, e_j \rangle = \delta_{ij}$ for $i, j \in \mathbb{N}$, where δ_{ij} denotes the Kronecker symbol. A linear operator (or, more simply, an operator) T in $\mathcal{H}$ is a linear map $u \mapsto Tu$ defined on a subspace $\mathcal{H}_0$ of $\mathcal{H}$, which is denoted by $D(T)$ and is called the domain of T. We denote by $R(T)$ the range of $\mathcal{H}_0$. We say that T is bounded if it is continuous from $D(T)$ (with the topology induced by the topology of $\mathcal{H}$) into

$\mathcal{H}$. When $D(T) = \mathcal{H}$, we recover the notion of linear continuous operators on $\mathcal{H}$. We denote by $\mathcal{L}(\mathcal{H})$ the space of continuous linear operators on $\mathcal{H}$ and recall that with

$$\|T\|_{\mathcal{L}(\mathcal{H})} = \sup_{u \neq 0} \frac{\|Tu\|_{\mathcal{H}}}{\|u\|_{\mathcal{H}}}, \tag{2.1.1}$$

$\mathcal{L}(\mathcal{H})$ is a Banach space. When $D(T)$ is not equal to $\mathcal{H}$, we shall always assume that

$$D(T) \text{ is dense in } \mathcal{H}. \tag{2.1.2}$$

Note that if in addition T is bounded, then it admits, by a standard unique extension theorem on complete metric spaces, a unique continuous extension to $\mathcal{H}$. In this case, the generalized notion is not interesting.

We are interested mainly in extensions of this theory and would like to consider unbounded operators. When we use this word, we mean more precisely "operators that are not necessarily bounded." The point is to find a natural notion to replace this notion of boundedness. This is the subject of the next definition.

An operator is called **closed** if the graph $G(T)$ of T is closed in $\mathcal{H} \times \mathcal{H}$. We recall that

$$G(T) := \{(x, y) \in \mathcal{H} \times \mathcal{H}, \, x \in D(T), \, y = Tx\}. \tag{2.1.3}$$

Equivalently, we can write the following.

Definition 2.1 (Closed operators.) Let T be an operator on $\mathcal{H}$ with (dense) domain $D(T)$. T is said to be closed if the conditions

- $u_n \in D(T)$,
- $u_n \to u$ in $\mathcal{H}$,
- $T u_n \to v$ in $\mathcal{H}$

imply

- $u \in D(T)$,
- $v = Tu$.

Sobolev spaces and Fourier transform   To treat some basic examples, we recall that the Sobolev space $H^s(\mathbb{R}^m)$ is defined as the space

$$H^s(\mathbb{R}^m) := \{u \in \mathcal{S}'(\mathbb{R}^m) \mid (1 + |\xi|^2)^{s/2} \, \hat{u} \in L^2(\mathbb{R}^m)\}. \tag{2.1.4}$$

Here, $\mathcal{S}'(\mathbb{R}^m)$ is the set of tempered distributions, and $\hat{u}$ (or $\mathcal{F}u$) denotes the Fourier transform in the sense of $\mathcal{S}'(\mathbb{R}^m)$, which for more regular distributions, say for $u \in \mathcal{S}(\mathbb{R}^m)$, takes the form

$$\hat{u}(\xi) = (2\pi)^{-m/2} \int e^{-ix \cdot \xi} u(x) \, dx. \tag{2.1.5}$$

We recall that $\mathcal{F}$ is a bijection from $\mathcal{S}(\mathbb{R}^m)$ onto $\mathcal{S}(\mathbb{R}^m)$, from $L^2(\mathbb{R}^m)$ onto $L^2(\mathbb{R}^m)$, and from $\mathcal{S}'(\mathbb{R}^m)$ onto $\mathcal{S}'(\mathbb{R}^m)$, and that its inverse is given by

$$\mathcal{F}^{-1}v(x) = (2\pi)^{-m/2} \int e^{ix \cdot \xi} v(\xi) \, d\xi. \tag{2.1.6}$$

Moreover, $\mathcal{F}$ is an isometry from $L^2(\mathbb{R}^m)$ onto $L^2(\mathbb{R}^m)$. This implies in particular the so-called Plancherel formula,

$$\int |\hat{u}(\xi)|^2 \, d\xi = \int |u(x)|^2 \, dx, \ \forall u \in L^2(\mathbb{R}^m). \tag{2.1.7}$$

$H^s(\mathbb{R}^m)$ is equipped with the natural Hilbertian norm:

$$||u||_{H^s}^2 := \int_{\mathbb{R}^m} (1 + |\xi|^2)^s |\hat{u}(\xi)|^2 \, d\xi.$$

By a Hilbertian norm, we mean that the norm is associated with a scalar product. When $s \in \mathbb{N}$, we can also describe H^s by

$$H^s(\mathbb{R}^m) := \{u \in L^2(\mathbb{R}^m) \mid D_x^\alpha u \in L^2, \ \forall \alpha \text{ s.t. } |\alpha| \leq s\}. \tag{2.1.8}$$

The natural norm associated with the second definition is equivalent to the first one.

Example 2.2

1. $T_0 = -\Delta$ with $D(T_0) = C_0^\infty(\mathbb{R}^m)$ is not closed.

   For this, it is enough to start from some u in $H^2(\mathbb{R}^m)$ that is not in $C_0^\infty(\mathbb{R}^m)$. Using the density of $C_0^\infty(\mathbb{R}^m)$ in $H^2(\mathbb{R}^m)$, there exists a sequence $u_n \in C_0^\infty(\mathbb{R}^m)$ such that $u_n \to u$ in $H^2(\mathbb{R}^m)$. The sequence $(u_n, -\Delta u_n)$ is contained in $G(T_0)$ and converges in $L^2 \times L^2$ to $(u, -\Delta u)$, which does not belong to $G(T_0)$.

2. $T_1 = -\Delta$ with $D(T_1) = H^2(\mathbb{R}^m)$ is closed.

   We observe that if $u_n \to u$ in L^2 and $(-\Delta u_n) \to v$ in L^2, then $-\Delta u = v \in L^2$. The last step is to observe that this implies that $u \in H^2(\mathbb{R}^m)$ (take the Fourier transform) and $(u, -\Delta u) \in G(T_1)$.

This example suggests another definition.

Definition 2.3 An operator T is called **closable** if the closure of the graph of T is a graph.

We can then define the closure $\overline{T}$ of the operator by a limiting procedure via its graph. We observe that we can consider

$$D(\overline{T}) := \{x \in \mathcal{H} \mid \exists y \text{ s.t. } (x, y) \in \overline{G(T)}\}.$$

For any $x \in D(\overline{T})$, the assumption that $\overline{G(T)}$ is a graph implies that y is unique. Consequently, we can define $\overline{T}$ by

$$\overline{T}x = y.$$

In other words, $\overline{T}$ is the operator whose graph is $\overline{G(T)}$ and, in a more explicit way, the domain of $\overline{T}$ is the set of the $x \in \mathcal{H}$ such that there exists a sequence (x_n) in $D(T)$ such that $x_n \to x \in \mathcal{H}$ and $T x_n$ is a Cauchy sequence. For such x, we can then define $\overline{T}x$ by

$$\overline{T}x = \lim_{n \to +\infty} T x_n.$$

$\overline{T}$ is called the closure of T.

Example 2.4 The Laplacian $T_0 = -\Delta$ with $D(T_0) = C_0^\infty(\mathbb{R}^m)$ is closable, and its closure is T_1.

Let us prove this, as an exercise. Let $u \in L^2(\mathbb{R}^m)$ be such that there exists $u_n \in C_0^\infty$ such that $u_n \to u$ in L^2 and $-\Delta u_n \to v$ in L^2. We obtain, by distribution theory, the result that $u \in L^2$ satisfies $-\Delta u = v \in L^2$. Hence, for a given u, v is unique and T_0 is closable. Let $\overline{T_0}$ be the closure of T_0 and $u \in D(\overline{T_0})$. The previous argument shows that $\Delta u \in L^2$. By the ellipticity of the Laplacian (use the Fourier transform), we obtain the result that $u \in H^2(\mathbb{R}^m)$. Consequently, we have shown that $D(\overline{T_0}) \subset H^2$. But $C_0^\infty(\mathbb{R}^m)$ is dense in H^2, and this gives the inverse inclusion, $H^2 \subset D(\overline{T_0})$. We have, consequently,

$$H^2 = D(T_1) = D(\overline{T_0}),$$

and it is then easy to verify that $T_1 = \overline{T_0}$.

These examples lead to a more general question.

Realization of differential operators as unbounded operators   Let $\Omega \subset \mathbb{R}^m$, and let $P(x, D_x)$ be a partial differential operator with C^∞ coefficients in Ω. Then the operator P^Ω defined by

$$D(P^\Omega) = C_0^\infty(\Omega), \ P^\Omega u = P(x, D_x)u, \forall u \in C_0^\infty(\Omega),$$

is closable. Here, $\mathcal{H} = L^2(\Omega)$. We have in fact

$$\overline{G(P^\Omega)} \subset \tilde{G}_\Omega := \{(u, f) \in \mathcal{H} \times \mathcal{H} \mid P(x, D_x)u = f \text{ in } \mathcal{D}'(\Omega)\}.$$

The proof is then a simple exercise in distribution theory. We just observe that if u_n is a sequence in $C_0^\infty(\Omega)$ such that u_n converges to u in L^2, then $P(x, D_x)u_n$ converges to $P(x, D_x)u$ in $\mathcal{D}'(\Omega)$. This inclusion shows that $\overline{G(P^\Omega)}$ is a graph. Note that the corresponding operator is defined as $P_{\min}^\Omega$, with domain

$$D(P_{\min}^\Omega) = \left\{ u \in L^2(\Omega) \mid \exists \text{ a sequence } u_n \in C_0^\infty(\Omega) \right.$$
$$\left. \text{s.t.} \begin{cases} u_n \to u \text{ in } L^2(\Omega) \\ P(x, D_x)u_n \text{ converges in } L^2(\Omega) \end{cases} \right\}.$$

The operator $P_{\min}^\Omega$ is then defined for such u by

$$P_{\min}^\Omega u = \lim_{n \to +\infty} P(x, D_x)u_n.$$

Using the theory of distributions, this gives

$$P_{\min}^\Omega u = P(x, D_x)u.$$

Note that there exists also a natural closed operator whose graph is $\tilde{G}_\Omega$ and which extends P^Ω. This is the operator $\tilde{P}^\Omega$, with domain

$$\tilde{D}^\Omega := \{u \in L^2(\Omega), \ P(x, D_x)u \in L^2(\Omega)\},$$

and such that

$$\tilde{P}^\Omega u = P(x, D_x)u, \forall u \in \tilde{D}^\Omega,$$

where the last equality is in the distributional sense. Note that $\tilde{P}^\Omega$ is an extension of $P_{\min}^\Omega$ in the sense that

$$\tilde{P}^\Omega u = P_{\min}^\Omega u, \ \forall u \in D(P_{\min}^\Omega).$$

Conclusion We have associated three natural operators with a differential operator $P(x, D_x)$ in an open set Ω. It is important to know the connection between these three operators better.

Remark 2.5 (Link between continuity and closedness.) If $\mathcal{H}_0 = \mathcal{H}$, the closed graph theorem states that a closed operator T is continuous.

2.2 Adjoints

When we have an operator T in $\mathcal{L}(\mathcal{H})$, it is easy to define the Hilbertian adjoint T^* by the identity

$$\langle u, T^*v \rangle_{\mathcal{H}} = \langle Tu, v \rangle_{\mathcal{H}}, \forall u \in \mathcal{H}, \forall v \in \mathcal{H}. \tag{2.2.1}$$

More precisely, the map $u \mapsto \langle Tu, v \rangle_{\mathcal{H}}$ defines a continuous linear map on $\mathcal{H}$ and can be expressed, using Riesz's theorem, which will be recalled in Section 3.1, as the scalar product with an element called T^*v. The linearity and the continuity of T^* are then easily proved using (2.2.1).

Let us now give a definition of the adjoint of an unbounded operator.

Definition 2.6 (Adjoint.) If T is an unbounded operator on $\mathcal{H}$ whose domain $D(T)$ is dense in $\mathcal{H}$, we first define the domain of T^* by

$$D(T^*) \quad = \{v \in \mathcal{H}, D(T) \ni u \mapsto \langle Tu, v \rangle_{\mathcal{H}},$$
$$\text{can be extended as a linear continuous form on } \mathcal{H}\}.$$

Using Riesz's theorem, there exists $f \in \mathcal{H}$ such that

$$\langle u, f \rangle = \langle Tu, v \rangle_{\mathcal{H}}, \forall u \in D(T).$$

The uniqueness of f is a consequence of the density of $D(T)$ in $\mathcal{H}$, and we can then define T^*v by

$$T^*v = f.$$

Remark 2.7 If $D(T) = \mathcal{H}$ and T is bounded, then we recover the Hilbertian adjoint as T^*.

As an example of the computation of an adjoint, let us show that with the T_0 and T_1 introduced in Example 2.2, we have

$$T_0^* = T_1.$$

We first obtain

$$D(T_0^*) = \{u \in L^2(\mathbb{R}^m) \mid \text{the map } C_0^\infty(\mathbb{R}^m) \ni v \mapsto \langle u, -\Delta v \rangle,$$
$$\text{can be extended as an antilinear continuous form on } L^2(\mathbb{R}^m)\}.$$

We observe that

$$\langle u, -\Delta v \rangle_{L^2} = \int_{\mathbb{R}^m} u \, \overline{(-\Delta v)} \, dx = (-\Delta u)(\bar{v}).$$

The last equality just means that we are considering the distribution $(-\Delta u)$ on the test function $\bar{v}$. The condition appearing in the definition is that this distribution is in $L^2(\mathbb{R}^m)$. Returning to the definition of $D(T_0^*)$, we obtain

$$D(T_0^*) = \{u \in L^2(\mathbb{R}^m) \mid -\Delta u \in L^2(\mathbb{R}^m)\}.$$

But, as already seen, this gives

$$D(T_0^*) = H^2(\mathbb{R}^m), \quad T_0^* u = -\Delta u, \quad \forall u \in H^2(\mathbb{R}^m).$$

Proposition 2.8 *T^* is a closed operator.*

Proof Let (v_n) be a sequence in $D(T^*)$ such that $v_n \to v$ in $\mathcal{H}$ and $T^* v_n \to w^*$ in $\mathcal{H}$ for some pair (v, w^*). We would like to show that (v, w^*) belongs to the graph of T^*.

For all $u \in D(T)$, we have

$$\langle Tu, v \rangle = \lim_{n \to +\infty} \langle Tu, v_n \rangle = \lim_{n \to +\infty} \langle u, T^* v_n \rangle = \langle u, w^* \rangle. \tag{2.2.2}$$

Recalling the definition of $D(T^*)$, we obtain from (2.2.2) the result that $v \in D(T^*)$ and $T^* v = w^*$. This means that (v, w^*) belongs to the graph of T^*. $\square$

Proposition 2.9 *Let T be an operator in $\mathcal{H}$ with domain $D(T)$. Then the graph $G(T^*)$ of T^* can be characterized by*

$$G(T^*) = \{V(\overline{G(T)})\}^\perp, \tag{2.2.3}$$

where V is the unitary operator defined on $\mathcal{H} \times \mathcal{H}$ by

$$V\{u, v\} = \{v, -u\}. \tag{2.2.4}$$

Proof We just observe that for any $u \in D(T)$ and $(v, w^*) \in \mathcal{H} \times \mathcal{H}$, we have the identity

$$\langle V(u, Tu), (v, w^*) \rangle_{\mathcal{H} \times \mathcal{H}} = \langle Tu, v \rangle_{\mathcal{H}} - \langle u, w^* \rangle_{\mathcal{H}}.$$

The right-hand side vanishes for all $u \in D(T)$ if and only if $v \in D(T^*)$ and $w^* = T^* v$, that is, if (v, w^*) belongs to $G(T^*)$. The left-hand side vanishes for all $u \in D(T)$ if and only if (v, w^*) belongs to $V(G(T))^\perp$. Basic Hilbertian analysis, using the continuity of V and $V^{-1} = -V$, then shows that

$$\{V(G(T))\}^\perp = \{\overline{V(G(T))}\}^\perp = \{V(\overline{G(T)})\}^\perp. \qquad \square$$

We have not yet analyzed the question of the conditions that the domain of the adjoint is dense in $\mathcal{H}$ under. This will be clarified in the next theorem.

Theorem 2.10 *Let T be a closable operator. Then we have:*

1. $D(T^*)$ is dense in $\mathcal{H}$.
2. $T^{**} := (T^*)^* = \overline{T}$,

where we have denoted by $\overline{T}$ the operator whose graph is $\overline{G(T)}$.

Proof For the first point, let us assume **in contradiction** that $D(T^*)$ is not dense in $\mathcal{H}$. Then there exists a $w \neq 0$ such that w is orthogonal to $\overline{D(T^*)}$. Consequently, we obtain the result that for any $v \in D(T^*)$, we have

$$\langle (0, w), \ (T^*v, -v) \rangle_{\mathcal{H} \times \mathcal{H}} = 0.$$

This shows that $(0, w)$ is orthogonal to $V(G(T^*))$.
    But the previous proposition gives

$$V(\overline{G(T)}) = G(T^*)^{\perp}.$$

We now apply V to this identity and obtain, using $V^2 = -I$,

$$V\left(G(T^*)^{\perp} \right) = \overline{G(T)}.$$

But, for any subspace $\mathcal{M} \subset \mathcal{H} \times \mathcal{H}$, we have

$$V(\mathcal{M}^{\perp}) = [V(\mathcal{M})]^{\perp},$$

as a consequence of the identity

$$\langle V(u, v), \ (x, y) \rangle_{\mathcal{H} \times \mathcal{H}} = -\langle (u, v), \ V(x, y) \rangle_{\mathcal{H} \times \mathcal{H}}.$$

Finally, we obtain the result that $(0, w)$ belongs to the closure of the graph of T, that is, the graph of $\overline{T}$, because T is closable and, consequently, $w = 0$. **This gives the contradiction**.
    For the second point, we first observe that since $D(T^*)$ is dense in $\mathcal{H}$, we can of course define $(T^*)^*$. Using again the previous proposition and the

closedness of T^*, we obtain $G(T^{**}) = \overline{G(T)}$ and $T^{**} = \overline{T}$. This means, more explicitly, that

$$D(T^{**}) = D(\overline{T}), \; T^{**}u = \overline{T}u, \; \forall u \in D(\overline{T}). \qquad \square$$

2.3 Symmetric and self-adjoint operators

Definition 2.11 (Symmetric operators.) We say that $T : \mathcal{H}_0 \mapsto \mathcal{H}$ is **symmetric** if it satisfies

$$\langle Tu, v \rangle_{\mathcal{H}} = \langle u, Tv \rangle_{\mathcal{H}}, \forall u, v \in \mathcal{H}_0.$$

Example 2.12 $T_0 = -\Delta$ with $D(T_0) = C_0^{\infty}(\mathbb{R}^m)$.

If T is symmetric, it is easy to see that

$$D(T) \subset D(T^*) \qquad (2.3.1)$$

and that

$$Tu = T^*u, \; \forall u \in D(T). \qquad (2.3.2)$$

The two conditions (2.3.1) and (2.3.2) express the property that $(T^*, D(T^*))$ is an extension of $(T, D(T))$.

Lemma 2.13 *A symmetric operator is closable.*

Proof It is enough to show that if u_n is a sequence in $D(T)$ such that, for some $\ell \in \mathcal{H}$, we have $u_n \to 0$ and $Tu_n \to \ell$, then $\ell = 0$. But this can be seen immediately by writing, for any $v \in D(T)$,

$$\langle Tu_n, v \rangle = \langle u_n, Tv \rangle,$$

and taking the limit $n \to +\infty$. This implies $\langle \ell, v \rangle = 0$ for all $v \in D(T)$ and, using the density of $D(T)$, we obtain the conclusion of the lemma. $\qquad \square$

Remark 2.14 This proof shows in fact that if $D(T^*)$ is dense, then T is closable, and we have already shown that the converse is true.

For a symmetric operator, we have consequently two natural closed extensions:

- The **minimal** one, denoted by $T_{\min}$ (or, previously, $\overline{T}$), which is obtained by taking the operator whose graph is the closure of the graph of T.
- The **maximal** one, denoted by $T_{\max}$, which is the adjoint of T.

If T^{sa} is a self-adjoint extension of T, then T^{sa} is automatically an extension of $T_{\min}$ and admits[1] $T_{\max}$ as an extension.

Definition 2.15 We say that T is **self-adjoint** if $T^* = T$, i.e.,

$$D(T) = D(T^*) \quad \text{and} \quad Tu = T^*u, \quad \forall u \in D(T).$$

Starting from a symmetric operator, it is natural to ask about the existence and uniqueness of a self-adjoint extension. We shall see later that a natural way is to prove the equality of $T_{\min}$ and $T_{\max}$.

Proposition 2.16 *A self-adjoint operator is closed.*

This can be seen immediately, because T^* is closed.

Proposition 2.17 *Let T be an invertible self-adjoint operator. Then T^{-1} is also self-adjoint.*

By "invertible," we mean here that T admits an inverse T^{-1} from $R(T)$ into $D(T)$.

Let us show first that $R(T)$ is dense in $\mathcal{H}$. This is done by proving that $R(T)^{\perp} = \{0\}$. Let $w \in \mathcal{H}$ be such that $\langle Tu, w \rangle_{\mathcal{H}} = 0, \forall u \in D(T)$. Returning to the definition of T^*, this implies in particular that $w \in D(T^*)$ and $T^*w = 0$. But T is self-adjoint and injective, and this implies that $w = 0$. Consequently, we know that $D(T^{-1})$ is dense in $\mathcal{H}$.

Returning to the analysis of the corresponding graphs, it is now easy to show the second assertion by using Proposition 2.9.

Remark 2.18 If T is self-adjoint, $T + \lambda I$ is self-adjoint for any real λ.

2.4 Notes

The material in this chapter is very basic, and we do not claim any originality in the exposition. Some books where this material has been presented are those by Brézis [Br], Hislop and Sigal [HiSi], Huet [Hu] (in French), Kato [Ka], Lévy-Bruhl [Le-Br] (in French), Reed and Simon [RS-I], and Robert [Ro] (in French).

Although we have tried to recall the main definitions, we have also assumed in the treatment of many examples that the reader has some basic knowledge of distribution theory. This topic is also covered in many standard books, including those by Hörmander [Ho2] and Zuily [Zu]. We should also mention some

---

[1] Use Proposition 2.9.

books that are more oriented toward semiclassical analysis, namely those by the author [He2], Dimassi and Sjöstrand [DiSj], and Zworski [Zw2].

2.5 Exercises

Exercise 2.1 Consider the operator L_{00} on $\mathbb{R}^+$ with $D(L_{00}) = C_0^\infty(\mathbb{R}^+)$ and $L_{00} = -d^2/dx^2$. Determine the closure of L_{00}. Determine L_{00}^*.

Exercise 2.2 (Analysis of differential operators.) Give simple criteria in the case of operators with constant coefficients $\sum_{|\alpha| \leq p} a_\alpha \partial_x^\alpha$ for obtaining symmetric operators on $C_0^\infty(\mathbb{R}^m)$. In particular, verify that the operator $D_{x_j} = (1/i)\partial_{x_j}$ is symmetric.

Exercise 2.3 Let V be a C^∞ function on $\mathbb{R}^m$. Consider $S_0 := -\Delta + V$, with domain $D(S_0) := C_0^\infty(\mathbb{R}^m)$. Show that S_0 is symmetric if and only if V is real-valued.

Exercise 2.4 On the interval $]0, 1[$, consider the operator $L := -i\partial_x$, with domain $D(L) := \{u \in H^1(]0, 1[), u(1) = 0\}$. Show that the operator is closed. Is the operator symmetric? Determine L^*.

3
Representation theorems

Given a complex Hilbert space $\mathcal{H}$, it is quite important to identify the space of continuous linear forms on $\mathcal{H}$, $\mathcal{L}(\mathcal{H}, \mathbb{C}) = \mathcal{H}'$, with the Hilbert space itself. This is done via Riesz's theorem. It is also useful to play with two Hilbertian scalar products, and, surprisingly, this leads rather simply to nontrivial results. This will be presented as the Lax–Milgram theorem, which is sometimes given the more modest name "Lax–Milgram lemma."

3.1 Riesz's theorem

Theorem 3.1 *(Riesz's representation theorem.) Let $u \mapsto F(u)$ be a linear continuous form[1] on $\mathcal{H}$. Then there exists a unique $w \in \mathcal{H}$ such that*

$$F(u) = \langle u, w \rangle_{\mathcal{H}}, \ \forall u \in \mathcal{H}. \tag{3.1.1}$$

Moreover, $F \mapsto w$ is an antilinear bijection from $\mathcal{H}' = \mathcal{L}(H, \mathbb{C})$ onto $\mathcal{H}$, and

$$\|w\|_{\mathcal{H}} = \|F\|_{\mathcal{L}(\mathcal{H}, \mathbb{C})}.$$

There is a similar version for antilinear maps, as follows.

Theorem 3.2 *Let $u \mapsto F(u)$ be an antilinear continuous form on $\mathcal{H}$. Then there exists a unique $w \in \mathcal{H}$ such that*

$$F(u) = \langle w, u \rangle_{\mathcal{H}}, \ \forall u \in \mathcal{H}. \tag{3.1.2}$$

Moreover, $F \mapsto w$ is a linear bijection from $\mathcal{H}' = \mathcal{L}(\mathcal{H}, \mathbb{C})$ onto $\mathcal{H}$, and

$$\|w\|_{\mathcal{H}} = \|F\|_{\mathcal{L}(\mathcal{H}, \mathbb{C})}.$$

---

[1] Some other authors use the terminology "linear functional."

3.2 Lax–Milgram situation

Let us now consider a continuous sesquilinear form a defined on $V \times V$,

$$(u, v) \mapsto a(u, v).$$

By "sesquilinear" we mean that for any $v \in V$, $u \mapsto a(u, v)$ is linear, and that for any $u \in V$, $v \mapsto a(u, v)$ is antilinear. We recall that because of the sesquilinearity, the continuity can be expressed by the existence of C such that

$$|a(u, v)| \leq C \, ||u||_V \cdot ||v||_V, \forall u, v \in V. \tag{3.2.1}$$

We can immediately associate a linear map $A \in \mathcal{L}(V)$ with this, using Riesz's theorem, such that

$$a(u, v) = \langle Au, v \rangle_V. \tag{3.2.2}$$

Definition 3.3 (V-ellipticity.) We say that a is V-elliptic if there exists $\alpha > 0$ such that

$$|a(u, u)| \geq \alpha \, ||u||_V^2, \forall u \in V. \tag{3.2.3}$$

Theorem 3.4 *(Lax–Milgram theorem.) Let a be a continuous sesquilinear form on $V \times V$. If a is V-elliptic, then A, as defined in (3.2.2), is an isomorphism from V onto V.*

Proof The proof is in three steps.

Step 1: A is injective.

We obtain the following from (3.2.2) and (3.2.3):

$$|\langle Au, u \rangle_V| \geq \alpha ||u||_V^2, \forall u \in V. \tag{3.2.4}$$

Using the Cauchy–Schwarz inequality in the left-hand side, we first obtain

$$||Au||_V \cdot ||u||_V \geq \alpha ||u||_V^2, \forall u \in V,$$

and, consequently,

$$||Au||_V \geq \alpha ||u||_V, \forall u \in V. \tag{3.2.5}$$

This clearly gives the injectivity, but actually gives us more.

Step 2: $A(V)$ is dense in V.

Let us consider $u \in V$ such that $\langle Av, u \rangle_V = 0$, $\forall v \in V$. In particular, we can take $v = u$. This gives $a(u, u) = 0$ and $u = 0$ using (3.2.3).

Step 3: $R(A) := A(V)$ is closed in V.

Let v_n be a Cauchy sequence in $A(V)$ and let u_n be the sequence such that $Au_n = v_n$. But, using (3.2.5), we obtain the result that u_n is a Cauchy sequence, which is consequently convergent to some $u \in V$. But the sequence Au_n tends to Au by continuity, and this shows that $v_n \to v = Au$ and $v \in R(A)$.

Step 4: A^{-1} is continuous.

The three previous steps show that A is bijective. The continuity of A^{-1} is a consequence of (3.2.5) or of the Banach theorem. $\qquad\square$

Remark 3.5 Assume, for simplicity, that V is a real Hilbert space. Using the isomorphism $\mathcal{I}$ between V and V' given by Riesz's theorem, we obtain also a natural operator $\mathcal{A}$ from V onto V' such that

$$a(u, v) = (\mathcal{A}u)(v), \ \forall v \in V. \tag{3.2.6}$$

We have

$$\mathcal{A} = \mathcal{I} \circ A.$$

3.3 An alternative point of view: the triple $V, \mathcal{H}, V'$

We now consider two Hilbert spaces V and $\mathcal{H}$ such that

$$V \subset \mathcal{H}. \tag{3.3.1}$$

By this notation for inclusion, we mean also that the injection of V into $\mathcal{H}$ is continuous or, equivalently, that there exists a constant $C > 0$ such that, for any $u \in V$, we have

$$\|u\|_{\mathcal{H}} \leq C \|u\|_V.$$

We also assume that

$$V \text{ is dense in } \mathcal{H}. \tag{3.3.2}$$

In this case, there exists a natural injection from $\mathcal{H}$ into the space V', which is defined as the space of continuous linear forms on V. We observe in fact that if $h \in \mathcal{H}$, then $V \ni u \mapsto \langle u, h \rangle_{\mathcal{H}}$ is continuous on V. So, there exists $\ell_h \in V'$ such that

$$\ell_h(u) = \langle u, h \rangle_{\mathcal{H}}, \forall u \in V.$$

The injectivity is a consequence of the density of V in $\mathcal{H}$.

Under the assumption that the sesquilinear form a is V-elliptic, we can also associate an unbounded operator S on $\mathcal{H}$ with a in the following way. We first define $D(S)$ by

$$D(S) = \{u \in V \mid v \mapsto a(u, v)$$

$$\text{is continuous on } V \text{ for the topology induced by } \mathcal{H}\}. \qquad (3.3.3)$$

Using again Riesz's theorem and the assumption (3.3.2), we can then define Su in $\mathcal{H}$ by

$$a(u, v) = \langle Su, v \rangle_{\mathcal{H}}, \ \forall v \in V. \qquad (3.3.4)$$

Theorem 3.4 is completed by the following theorem.

Theorem 3.6 *Let a be a continuous sesquilinear form on $V \times V$. If a is V-elliptic and (3.3.1) and (3.3.2) are true, then S, as defined in (3.3.3)–(3.3.4), is bijective from $D(S)$ onto $\mathcal{H}$ and $S^{-1} \in \mathcal{L}(\mathcal{H})$.*
Moreover, $D(S)$ is dense in $\mathcal{H}$.

Proof We first show that S is injective. This is a consequence of

$$\alpha||u||_{\mathcal{H}}^2 \leq C\,\alpha||u||_V^2 \leq C|a(u, u)| = C\,|\langle Su, u \rangle_{\mathcal{H}}| \leq C\,||Su||_{\mathcal{H}} \cdot ||u||_{\mathcal{H}},$$

$$\forall u \in D(S),$$

which leads to

$$\alpha||u||_{\mathcal{H}} \leq C\,||Su||_{\mathcal{H}}, \ \forall u \in D(S). \qquad (3.3.5)$$

We obtain the surjectivity directly in the following way. If $h \in \mathcal{H}$ and if $w \in V$ is chosen such that

$$\langle h, v \rangle_{\mathcal{H}} = \langle w, v \rangle_V, \forall v \in V$$

(the existence of w follows from Riesz's theorem), then we can take $u = A^{-1}w$ in V, which is a solution of

$$a(u, v) = \langle w, v \rangle_V.$$

We can then show that $u \in D(S)$, using the identity

$$a(u, v) = \langle h, v \rangle_{\mathcal{H}}, \ \forall v \in V,$$

and obtain simultaneously

$$Su = h.$$

The continuity of S^{-1} is a consequence of (3.3.5).

We now show the last statement of the theorem, i.e., the density of $D(S)$ in $\mathcal{H}$. Let $h \in \mathcal{H}$ such that

$$\langle u, h \rangle_{\mathcal{H}} = 0, \, \forall u \in D(S).$$

By the surjectivity of S, there exists $v \in D(S)$ such that

$$Sv = h.$$

We obtain

$$\langle Sv, u \rangle_{\mathcal{H}} = 0, \, \forall u \in D(S).$$

Taking $u = v$ and using the V-ellipticity, we obtain the result that $v = 0$ and, consequently, $h = 0$.   □

The Hermitian case   We now consider a Hermitian sesquilinear form, that is, a form satisfying

$$a(u, v) = \overline{a(v, u)}, \, \forall u, v \in V. \tag{3.3.6}$$

In this case we obtain additional properties.

Theorem 3.7 *If a is Hermitian and V-elliptic, we have:*

1. *S is closed.*
2. *$S = S^*$.*
3. *$D(S)$ is dense in V.*

Proof of 2   We first observe that the assumption of Hermiticity gives

$$\langle Su, v \rangle_{\mathcal{H}} = \langle u, Sv \rangle_{\mathcal{H}}, \, \forall u \in D(S), \, \forall v \in D(S). \tag{3.3.7}$$

In other words, S is symmetric. This means in particular that

$$D(S) \subset D(S^*). \tag{3.3.8}$$

Let $v \in D(S^*)$. Using the surjectivity of S, there exists $v_0 \in D(S)$ such that

$$Sv_0 = S^*v.$$

For all $u \in D(S)$, we obtain the result that

$$\langle Su, v_0 \rangle_{\mathcal{H}} = \langle u, Sv_0 \rangle_{\mathcal{H}} = \langle u, S^*v \rangle_{\mathcal{H}} = \langle Su, v \rangle_{\mathcal{H}}.$$

Using again the surjectivity of S, we obtain $v = v_0 \in D(S)$. This shows that $D(S) = D(S^*)$ and $Sv = S^*v$, $\forall v \in D(S)$. $\qquad\square$

Proof of 1 S is closed because S^* is closed and $S = S^*$. $\qquad\square$

Proof of 3 Let $h \in V$ be such that

$$\langle u, h \rangle_V = 0, \ \forall u \in D(S).$$

Let $f \in V$ be such that $Af = h$ (A is an isomorphism from V onto V). We then have

$$0 = \langle u, h \rangle_V = \langle u, Af \rangle_V = \overline{\langle Af, u \rangle_V} = \overline{a(f, u)} = a(u, f) = \langle Su, f \rangle_{\mathcal{H}}.$$

Using the surjectivity, we obtain $f = 0$ and, consequently, $h = 0$. $\qquad\square$

Remark 3.8 Theorem 3.7 gives us a rather easy way to construct self-adjoint operators. This will be combined with some completion arguments in the next chapter to obtain the Friedrichs extension.

3.4 Notes

We have assumed that the reader knows about this material, but have recalled it for completeness. Here we have followed, rather closely, one chapter of Huet's book [Hu]. The Lax–Milgram version with the triple $V, \mathcal{H}, V'$ is due to Lions [Lio1, Lio2]. The V-ellipticity is also called the V-coercivity or coercivity in the literature.

3.5 Exercises

Exercise 3.1 Let Ω be a bounded regular open set in $\mathbb{R}^n$ and let ρ be a continuous positive function, and consider the sesquilinear form

$$(u, v) \mapsto \int_{\Omega} \nabla u \cdot \nabla \bar{v} \, \rho(x) \, dx + \int_{\Omega} u(x) \bar{v}(x) \, dx.$$

Show that this is V-elliptic on a suitable V and that the results of Section 3.3 can be used for some triple $(V, \mathcal{H}, V')$. Describe the corresponding S.

Exercise 3.2 Consider the quadratic form

$$u \mapsto \int_0^{+\infty} |u'(t)|^2 \, dt + \int_0^{+\infty} |u(t)|^2 \, dt + \gamma |u(0)|^2$$

on $H^1(\mathbb{R}^+)$, where $\gamma \neq 0$.

(a) Write the corresponding sesquilinear form, and give optimal conditions on γ under which all the assumptions of the Lax–Milgram theorem are satisfied.

(b) Describe the associated operator S (and, in particular, its domain).

(c) Under what conditions on γ does S have eigenvalues?

4

Semibounded operators and the Friedrichs extension

Many symmetric operators are semibounded, and in this case there is a natural way to define a self-adjoint extension by application of a version of the Lax–Milgram theorem, via a completion argument.

4.1 Definition and basic example

Definition 4.1 Let T_0 be a symmetric (see Definition 2.11) unbounded operator with domain $D(T_0)$. We say that T_0 is semibounded (from below) if there exists a constant C such that

$$\langle T_0 u, u \rangle_{\mathcal{H}} \geq -C \|u\|_{\mathcal{H}}^2, \ \forall u \in D(T_0). \tag{4.1.1}$$

Example 4.2 (The Schrödinger operator.) We consider on $\mathbb{R}^m$ the operator

$$P_V(x, D_x) := -\Delta + V(x), \tag{4.1.2}$$

where $V(x)$ is a continuous real function on $\mathbb{R}^m$ (called the potential) such that there exists C such that

$$V(x) \geq -C, \ \forall x \in \mathbb{R}^m. \tag{4.1.3}$$

Then the operator T_0 defined by

$$D(T_0) = C_0^\infty(\mathbb{R}^m) \text{ and } T_0 u = P_V(x, D_x)u, \ \forall u \in D(T_0),$$

is a symmetric, semibounded operator.

We have in fact, with $\mathcal{H} = L^2(\mathbb{R}^m)$,

$$
\begin{aligned}
\langle P(x, D_x)u, u \rangle_{\mathcal{H}} &= \int_{\mathbb{R}^m} (-\Delta u + Vu) \cdot \bar{u} \, dx \\
&= \int_{\mathbb{R}^m} |\nabla u(x)|^2 \, dx + \int_{\mathbb{R}^m} V(x)|u(x)|^2 \, dx \\
&\geq -C \|u\|_{\mathcal{H}}^2, \tag{4.1.4}
\end{aligned}
$$

for any $u \in C_0^\infty(\mathbb{R}^m)$.

This is of course a basic example. But a natural question immediately occurs. Is it necessary that the potential V is semibounded? There are indeed natural nonsemibounded potentials for which one may hope for semiboundedness of the associated quantum Hamiltonian $-\Delta + V$. The first example will be analyzed in the next section.

4.2 Analysis of the Coulomb case

By the Coulomb case, we mean the analysis on $\mathbb{R}^3$ of the Schrödinger operator

$$S_Z := -\Delta - \frac{Z}{r}$$

(with $r = \sqrt{x^2 + y^2 + z^2}$ and $Z \in \mathbb{R}$ (the charge)) or of the Klein–Gordon operator

$$K_Z := \sqrt{-\Delta + 1} - \frac{Z}{r}.$$

Note that $1/r \in L^2_{\mathrm{loc}}(\mathbb{R}^3)$, and this allows us to define S_Z as an unbounded symmetric operator on $L^2(\mathbb{R}^3)$ with domain $C_0^\infty(\mathbb{R}^3)$. Note also that the operator $\sqrt{-\Delta + 1}$ can easily be defined on $C_0^\infty(\mathbb{R}^3)$ (and, more generally, on $\mathcal{S}(\mathbb{R}^3)$), using the Fourier transform $\mathcal{F}$ introduced in (2.1.5), by

$$\mathcal{F}(\sqrt{-\Delta + 1}\, u)(p) = \sqrt{p^2 + 1}\, (\mathcal{F}u)(p).$$

There are two inequalities that play an important role in atomic physics. The first is the Hardy inequality,

$$\int_{\mathbb{R}^3} |x|^{-2} |u(x)|^2\, dx \leq 4 \int_{\mathbb{R}^3} |p|^2\, |\hat{u}(p)|^2\, dp, \qquad (4.2.1)$$

and the second is due to Kato and states that

$$\int_{\mathbb{R}^3} |x|^{-1} |u(x)|^2\, dx \leq \frac{\pi}{2} \int_{\mathbb{R}^3} |p|\, |\hat{u}(p)|^2\, dp. \qquad (4.2.2)$$

One proof of the Hardy inequality consists in writing that, for any $\gamma \in \mathbb{R}$ and any $u \in C_0^\infty(\mathbb{R}^3; \mathbb{R})$, we have

$$\int_{\mathbb{R}^3} \left| \nabla u + \gamma \frac{x}{|x|^2} u \right|^2 dx \geq 0. \qquad (4.2.3)$$

This leads to

$$\int_{\mathbb{R}^3} |\nabla u|^2 + \gamma^2 \frac{1}{|x|^2} |u|^2\, dx \geq -2\gamma \int_{\mathbb{R}^3} \nabla u \cdot \frac{x}{|x|^2} u\, dx.$$

But an integration by parts gives

$$-2 \int_{\mathbb{R}^3} \nabla u \cdot \frac{x}{|x|^2} u \, dx = \int_{\mathbb{R}^3} |u(x)|^2 \left(\text{div} \left(\frac{x}{|x|^2} \right) \right) dx$$

$$= \int_{\mathbb{R}^3} |u(x)|^2 \left(\frac{1}{|x|^2} \right) dx.$$

Optimizing with respect to γ leads to $\gamma = \frac{1}{2}$ and then gives the result.

Remark 4.3 The same idea works for the problem on $\mathbb{R}^m$ ($m \geq 3$) but fails for $m = 2$. So, a good exercise[1] is to look for substitutes in this case by starting from the inequality

$$\int_{\mathbb{R}^3} \left| \nabla u - \gamma(x) \frac{x}{|x|^2} u(x) \right|^2 dx \geq 0. \tag{4.2.4}$$

The function $\gamma(x)$ can be assumed to be radial, i.e., $\gamma(x) = g(|x|)$, and the question is to find a differential inequality on g leading to a weaker Hardy-type inequality. One may, for example, try $g(r) = \ln(r)$.

For the proof of the inequality (4.2.2), there is a another nice and tricky estimate, which, as far as we know, goes back to Hardy and Littlewood. In the case of the Coulomb potential, we can write, using the explicit computation of the Fourier transform of $x \mapsto 1/|x|$,

$$\int_{\mathbb{R}^3} \int_{\mathbb{R}^3} \widehat{u}(p) \frac{1}{|p - p'|^2} \overline{\widehat{u}}(p') \, dp \, dp'$$

$$= \int_{\mathbb{R}^3 \times \mathbb{R}^3} \widehat{u}(p) \frac{h(p)}{h(p')} \cdot \frac{h(p')}{h(p)} \frac{1}{|p - p'|^2} \overline{\widehat{u}}(p') \, dp \, dp'$$

$$= \int_{\mathbb{R}^3 \times \mathbb{R}^3} \frac{1}{|p - p'|} \widehat{u}(p) \frac{h(p)}{h(p')} \cdot \frac{h(p')}{h(p)} \frac{1}{|p - p'|} \overline{\widehat{u}}(p') \, dp \, dp',$$

where h is a positive measurable function to be determined later. We then use the Cauchy–Schwarz inequality in the last equality in order to obtain

$$\left| \int \int \widehat{u}(p) \frac{1}{|p - p'|^2} \overline{\widehat{u}}(p') dp \cdot dp' \right|$$

$$\leq \left(\int |\widehat{u}(p)|^2 \left| \frac{h(p)}{h(p')} \right|^2 \frac{1}{|p - p'|^2} dp' \, dp \right)^{1/2}$$

---

[1] We thank M. J. Esteban for explaining this trick to us.

$$\times \left(\int |\widehat{u}(p')|^2 \left| \frac{h(p')}{h(p)} \right|^2 \frac{1}{|p - p'|^2} dp' \, dp \right)^{1/2}$$

$$= \int |\widehat{u}(p)|^2 \left(\int \left| \frac{h(p)}{h(p')} \right|^2 \frac{1}{|p - p'|^2} dp' \right) dp$$

$$= \int h(p)^2 |\widehat{u}(p)|^2 \left(\int \left| \frac{1}{h(p')} \right|^2 \frac{1}{|p - p'|^2} dp' \right) dp.$$

We now write $p' = \omega'|p|$ in the integral

$$\int \left| \frac{1}{h(p')} \right|^2 \frac{1}{|p - p'|^2} dp'.$$

We then take

$$h(p) = |p|.$$

The integral becomes

$$\int \left| \frac{1}{h(p')} \right|^2 \frac{1}{|p - p'|^2} dp' = |p|^{-1} \int \frac{1}{|\omega'|^2 \, |\omega - \omega'|^2} d\omega',$$

with $p = \omega |p|$. This is clearly a convergent integral. Moreover, observing the invariance with respect to rotation, it can be shown that the integral is independent of $\omega \in \mathbb{S}^2$. Hence we can compute it with $\omega = (0, 0, 1)$.

Finally, we obtain the existence of an explicit constant C such that

$$\left| \int \int \widehat{u}(p) \frac{1}{|p - p'|^2} \overline{\widehat{u}}(p') \, dp \, dp' \right| \leq C \int_{\mathbb{R}^3} |p| \, |\widehat{u}(p)|^2 \, dp.$$

Optimization of the trick leads to $C = \pi^3$.

Let us now show how these inequalities can be used.

If we use (4.2.1), we obtain the semiboundedness of the Schrödinger Coulomb operator for any $Z > 0$ (using the Cauchy–Schwarz inequality):

$$\int_{\mathbb{R}^3} \frac{1}{r} |u(x)|^2 \, dx \leq \left(\int \frac{1}{r^2} |u|^2 dx \right)^{1/2} \cdot \|u\|.$$

But we can rewrite the Hardy inequality in the form

$$\int_{\mathbb{R}^3} \frac{1}{r^2} |u(x)|^2 \, dx \leq 4 \langle -\Delta u, u \rangle_{L^2(\mathbb{R}^3)}.$$

So we obtain, for any $\epsilon > 0$,

$$\int_{\mathbb{R}^3} \frac{1}{r} |u|^2 \, dx \leq \epsilon \langle -\Delta u, u \rangle_{L^2} + \frac{1}{\epsilon} ||u||^2. \qquad (4.2.5)$$

This leads to

$$\langle S_Z u, u \rangle_{L^2} \geq (1 - \epsilon Z) \langle -\Delta u, u \rangle - \frac{Z}{\epsilon} ||u||^2.$$

Finally, taking $\epsilon = 1/Z$, we can show that

$$\langle S_Z u, u \rangle_{L^2} \geq -Z^2 ||u||^2. \qquad (4.2.6)$$

However, we are not optimal here. It can actually be proven (see any standard book on quantum mechanics) that the negative spectrum of this operator is discrete and is described by a sequence of eigenvalues tending to 0, namely $-Z^2/4n^2$ with $n \in \mathbb{N}^*$. An eigenfunction related to the lowest eigenvalue, $-\frac{1}{4}$ (for $Z = 1$), is given by $x \mapsto \exp -\frac{1}{2}|x|$. To prove this last statement, instead of using Hardy's inequality or the inequality (4.2.2), we can observe that

$$\left\| \nabla u - \rho \frac{x}{|x|} u \right\|^2 \geq 0,$$

and then take $\rho = -Z/2$.

There is another way to see that the behavior with respect to Z is correct. We just observe some invariance of the model. Let us suppose that we have proved the inequality for $Z = 1$. In order to treat the general case $Z \neq 0$, we make a change of variable $x = \rho y$. The operator S_Z becomes, in the new coordinates,

$$\tilde{S}_Z = \rho^{-2}(-\Delta_y) - \frac{Z}{\rho y}.$$

Taking $\rho = Z^{-1}$, we obtain

$$\tilde{S}_Z = Z^2 \left(-\Delta_y - \frac{1}{y} \right).$$

The other inequality (4.2.2) is much better in this respect and quite important for the analysis of the relativistic case. Let us see what we obtain in the case of the Klein–Gordon operator using this inequality. In this case we have

$$\langle K_Z u, u \rangle_{L^2} \geq \left(1 - Z\frac{\pi}{2} \right) \langle \sqrt{-\Delta + 1}\, u, u \rangle_{L^2}.$$

Here, the nature of the result is different. The proof tells us only that K_Z is semibounded if $Z \leq 2/\pi$. This is actually more than a technical problem!

4.3 The Friedrichs extension

Theorem 4.4 *A symmetric semibounded operator T_0 on $\mathcal{H}$ (with $D(T_0)$ dense in $\mathcal{H}$) admits a self-adjoint extension.*

The extension constructed in the proof of this theorem is called the Friedrichs extension. The proof can be seen as a variant of the Lax–Milgram theorem. We can assume, in fact, possibly with T_0 replaced by $T_0 + \lambda_0 Id$, that T_0 satisfies

$$\langle T_0 u, u \rangle_{\mathcal{H}} \geq ||u||_{\mathcal{H}}^2, \ \forall u \in D(T_0). \tag{4.3.1}$$

Let us consider the associated form initially defined on $D(T_0) \times D(T_0)$,

$$(u, v) \mapsto a_0(u, v) := \langle T_0 u, v \rangle_{\mathcal{H}}. \tag{4.3.2}$$

The inequality (4.3.1) states that

$$a_0(u, u) \geq ||u||_{\mathcal{H}}^2, \ \forall u \in D(T_0). \tag{4.3.3}$$

We introduce V as the completion in $\mathcal{H}$ of $D(T_0)$ for the norm

$$u \mapsto p_0(u) = \sqrt{a_0(u, u)}.$$

More concretely, $u \in \mathcal{H}$ belongs to V if there exists $u_n \in D(T_0)$ such that $u_n \rightarrow u$ in $\mathcal{H}$ and u_n is a Cauchy sequence for the norm p_0.

We obtain the following candidate natural norm for V:

$$||u||_V = \lim_{n \rightarrow +\infty} p_0(u_n), \tag{4.3.4}$$

where (u_n) is a Cauchy sequence for p_0 tending to u in $\mathcal{H}$.

We shall now show that the definition does not depend on the Cauchy sequence. This is the subject of the following lemma.

Lemma 4.5 *Let (x_n) be a Cauchy sequence in $D(T_0)$ for p_0 such that $x_n \rightarrow 0$ in $\mathcal{H}$. Then $p_0(x_n) \rightarrow 0$.*

Proof of the lemma First, we observe that $p_0(x_n)$ is a Cauchy sequence in $\mathbb{R}^+$ and, consequently, convergent in $\overline{\mathbb{R}^+}$.

Let us suppose, in contradiction, that

$$p_0(x_n) \to \alpha > 0. \tag{4.3.5}$$

We observe that

$$a_0(x_n, x_m) = a_0(x_n, x_n) + a_0(x_n, x_m - x_n), \tag{4.3.6}$$

and that a Cauchy–Schwarz inequality is satisfied,

$$|a_0(x_n, x_m - x_n)| \leq \sqrt{a_0(x_n, x_n)} \cdot \sqrt{a_0(x_m - x_n, x_m - x_n)}. \tag{4.3.7}$$

Using also the fact that x_n is a Cauchy sequence for p_0 and (4.3.6), we obtain the result that

$$\forall \epsilon > 0, \exists N \text{ s.t. } \forall n \geq N, \forall m \geq N, |a_0(x_n, x_m) - \alpha^2| \leq \epsilon. \tag{4.3.8}$$

We take $\epsilon = \alpha^2/2$ and consider the corresponding N given by (4.3.8). Returning to the definition of a_0, we obtain

$$|a_0(x_n, x_m)| = |\langle T_0 x_n, x_m \rangle| \geq \frac{1}{2}\alpha^2, \ \forall n \geq N, \forall m \geq N. \tag{4.3.9}$$

But as $m \to +\infty$, the left-hand side of (4.3.9) tends to 0 because, by assumption, $x_m \to 0$, and this gives a contradiction. $\qquad\square$

We now observe that

$$\|u\|_V \geq \|u\|_{\mathcal{H}}, \tag{4.3.10}$$

as a consequence of (4.3.3) and (4.3.4). This means that the injection of V in $\mathcal{H}$ is continuous. Note also that V, which contains $D(T_0)$, is dense in $\mathcal{H}$, by the density of $D(T_0)$ in $\mathcal{H}$. Moreover, we obtain a natural scalar product on V by extension of a_0:

$$\langle u, v \rangle_V := \lim_{n \to +\infty} a_0(u_n, v_n), \tag{4.3.11}$$

where u_n and v_n are Cauchy sequences for p_0 tending respectively to u and v in $\mathcal{H}$.

By the second version of the Lax–Milgram theorem (see Theorem 3.6) applied with

$$a(u, v) := \langle u, v \rangle_V,$$

we obtain an unbounded self-adjoint operator S on $\mathcal{H}$ extending T_0 whose domain $D(S)$ satisfies $D(S) \subset V$.

Remark 4.6 (Friedrichs extension starting from a sesquilinear form.) One can also start directly from a semibounded sesquilinear form a_0 defined on a dense subspace of $\mathcal{H}$. As we shall see below, this is actually the right way to proceed for the Neumann realization of the Laplacian, where we consider on $C^\infty(\overline{\Omega})$ the sequilinear form

$$(u, v) \mapsto \langle \nabla u, \nabla v \rangle.$$

4.4 Applications

The reader is assumed to have some minimal knowledge of Sobolev spaces and traces of distributions to read this section.

4.4.1 The Dirichlet realization

Let Ω be an open set in $\mathbb{R}^m$ such that $\overline{\Omega}$ is a compact set, and let T_0 be the unbounded operator defined by

$$D(T_0) = C_0^\infty(\Omega), \; T_0 = -\Delta.$$

The Hilbert space involved is $\mathcal{H} = L^2(\Omega)$. It is clear that T_0 is symmetric and nonnegative[2] (and hence semibounded). Following the previous general construction, we prefer to consider $\tilde{T}_0 := T_0 + Id$.

It is easy to see that V is the closure in $H^1(\Omega)$ of $C_0^\infty(\Omega)$. This means, at least if Ω is regular, the space $H_0^1(\Omega)$. We recall that there is another way of describing $H_0^1(\Omega)$. We can define it as the subspace in $H^1(\Omega)$ of the distributions whose trace is zero at the boundary $\Gamma = \partial\Omega$. More precisely, under the assumption that Γ is regular, there exists a unique application γ_0, continuous from $H^1(\Omega)$ onto $H^{1/2}(\Gamma)$, extending the map $C^\infty(\overline{\Omega}) \ni u \mapsto u_\Gamma$, where u_Γ is the restriction of u to Γ, which is called the trace. The domain of S is then described as

$$D(S) := \{u \in H_0^1(\Omega) \mid -\Delta u \in L^2(\Omega)\}.$$

S is then the operator $(-\Delta + 1)$ acting in the sense of distributions.

When Ω is regular, a standard regularity theorem for second-order elliptic operators permits one to show that

$$D(S) = H^2(\Omega) \cap H_0^1(\Omega). \tag{4.4.1}$$

So, we have shown the following theorem.

---

[2] We shall see later that it is in fact positive.

Theorem 4.7 *The operator T_1 defined by*

$$D(T_1) = H^2(\Omega) \cap H^1_0(\Omega), \ T_1 = -\Delta,$$

is self-adjoint; it is called the Dirichlet realization of $-\Delta$ in Ω.

We have just to observe that $T_1 = S - 1$ and use Remark 2.18.

Note that T_1 is a self-adjoint extension of T_0. In Section 4.4.4, we shall construct another self-adjoint extension of T_0. Hence, when Ω is relatively compact, we can construct two different self-adjoint extensions of T_0. We say in this case that T_0 is not essentially self-adjoint (see Chapter 9).

4.4.2 The harmonic oscillator

We can start from

$$H_0 = -\Delta + |x|^2 + 1,$$

with domain

$$D(H_0) = C^\infty_0(\mathbb{R}^m).$$

Following the scheme of the construction of the Friedrichs extension (see Section 4.3), we look first at the completion V in L^2 of C^∞_0 for the norm

$$u \mapsto \sqrt{\int_{\mathbb{R}^m} |\nabla u(x)|^2 \, dx + \int_{\mathbb{R}^m} |x|^2 |u(x)|^2 \, dx + \int_{\mathbb{R}^m} |u(x)|^2 \, dx}.$$

We obtain

$$V = B^1(\mathbb{R}^m) := \{u \in H^1(\mathbb{R}^m) \mid x_j u \in L^2(\mathbb{R}^m), \ \forall j \in [1, \cdots, m]\}.$$

One can indeed show that $V \subset B^1(\mathbb{R}^m)$ and then obtain the equality by proving the property that $C^\infty_0(\mathbb{R}^m)$ is dense in $B^1(\mathbb{R}^m)$. The domain of S can then be determined as

$$D(S) = \{u \in B^1(\mathbb{R}^m) \mid (-\Delta + |x|^2 + 1)u \in L^2(\mathbb{R}^m)\}.$$

We should mention that by a regularity theorem based on the method of differential quotients, it can actually be shown that

$$D(S) = B^2(\mathbb{R}^m) := \{u \in H^2(\mathbb{R}^m) \mid x^\alpha u \in L^2(\mathbb{R}^m), \ \forall \alpha \ \text{s.t.} \ |\alpha| \leq 2\}.$$

4.4.3 The Schrödinger operator with a Coulomb potential

Consequently, we start from

$$D(T_0) = C_0^\infty(\mathbb{R}^3), \ T_0 = -\Delta - \frac{1}{r}. \tag{4.4.2}$$

We have seen that T_0 is semibounded and, replacing T_0 by $T_0 + 2$, the assumptions of the proof of Friedrich's extension theorem are satisfied. We now claim that

$$V = H^1(\mathbb{R}^3).$$

Bearing in mind that $C_0^\infty(\mathbb{R}^3)$ is dense in $H^1(\mathbb{R}^3)$, we have just to verify that the norm p_0 and the norm $||\cdot||_{H^1(\mathbb{R}^3)}$ are equivalent on $C_0^\infty(\mathbb{R}^3)$. This results immediately from (4.2.5).

Hardy's inequality also shows that if $u \in H^1(\mathbb{R}^3)$, then $(1/r)u$ is in $L^2(\mathbb{R}^3)$. Returning to the equation, one obtains the result that if $u \in D(S)$, then $\Delta u \in L^2(\mathbb{R}^3)$, and hence $D(S) = H^2(\mathbb{R}^3)$.

4.4.4 The Neumann realization

Let Ω be a bounded domain with a regular boundary in $\mathbb{R}^m$. Take $\mathcal{H} = L^2(\Omega)$. Let us consider the sesquilinear form

$$a_0(u, v) = \int_\Omega \langle \nabla u, \nabla v \rangle_{\mathbb{C}^m} \, dx + \int_\Omega u\bar{v} dx,$$

on $C^\infty(\overline{\Omega})$.

Using Remark 4.6 and the density of $C^\infty(\overline{\Omega})$ in $H^1(\Omega)$, we can extend the sesquilinear form to $V = H^1(\Omega)$. According to the definition of the domain of S, we observe that for $u \in D(S)$, there should exist some $f \in L^2(\Omega)$ such that, for all $v \in H^1(\Omega)$,

$$a(u, v) = \int_\Omega f(x)\, \bar{v}(x)\, dx. \tag{4.4.3}$$

Then, by taking $v \in C_0^\infty(\Omega)$ in (4.4.3), one can show first that in the sense of distributions,

$$-\Delta u + u = f, \tag{4.4.4}$$

and consequently that

$$D(S) \subset W(\Omega) := \{u \in H^1(\Omega) \mid -\Delta u \in L^2(\Omega)\}. \tag{4.4.5}$$

But this is not enough to characterize the domain.

We first recall the Green–Riemann formula,

$$\int_\Omega \langle \nabla u, \nabla v \rangle_{\mathbb{C}^m} \, dx = \int_\Omega (-\Delta u) \cdot \bar{v} \, dx + \int_{\partial\Omega} (\partial u / \partial \nu) \bar{v} \, d\mu_{\partial\Omega}, \quad (4.4.6)$$

where $d\mu_{\partial\Omega}$ is the induced measure on the boundary, which is clearly true for $u \in H^2(\Omega)$ (or for $u \in C^1(\overline{\Omega})$) and $v \in H^1(\Omega)$. Unfortunately, we do not know that $W(\Omega) \subset H^2(\Omega)$, and the inclusion in $H^1(\Omega)$ is not sufficient for defining the normal trace. But this formula can be extended in the following way.

We first observe that for $v \in H_0^1(\Omega)$ and $u \in W$, we have

$$\int_\Omega \langle \nabla u \mid \nabla v \rangle_{\mathbb{C}^m} \, dx = \int_\Omega (-\Delta u) \cdot \bar{v} \, dx. \quad (4.4.7)$$

This shows that the expression

$$\Phi_u(v) := \int_\Omega \langle \nabla u \mid \nabla v \rangle \, dx - \int_\Omega (-\Delta u) \cdot \bar{v} \, dx,$$

which is well defined for $u \in W$ and $v \in H^1(\Omega)$, depends only on the restriction of v to $\partial\Omega$.

If $v_0 \in C^\infty(\partial\Omega)$, we can then extend v_0 inside Ω as a function $v = R v_0$ in $C^\infty(\overline{\Omega})$. More precisely, it is known that the trace map γ_0 from $H^1(\Omega)$ into $H^{1/2}(\partial\Omega)$ is surjective and admits a continuous right inverse R by choosing a more specific continuous right inverse from $H^{1/2}(\partial\Omega)$ into $H^1(\Omega)$. Hence the map $C^\infty(\partial\Omega) \ni v_0 \mapsto \Phi_u(R v_0)$ defines a distribution in $\mathcal{D}'(\partial\Omega)$ that can be extended as a continuous map on $H^{1/2}(\partial\Omega)$ (i.e., an element of $H^{-1/2}(\partial\Omega)$). Now we observe also that when $u \in C^1(\overline{\Omega})$ or $u \in H^2(\Omega)$, the Green–Riemann formula shows that

$$\Phi_u(v_0) = \int_{\partial\Omega} (\partial u / \partial \nu) \bar{v}_0 \, d\mu_{\partial\Omega}.$$

So, we have found a natural way to extend the notion of the trace of the normal derivative for $u \in W$, and we write

$$\gamma_1 u = \Phi_u.$$

We then conclude (using (4.4.4) and again (4.4.3), this time in full generality) that

$$D(S) = \{u \in W(\Omega) \mid \gamma_1 u = 0\},$$

and that

$$S = -\Delta + 1.$$

The operator S is called the Neumann realization of the Laplacian in $L^2(\Omega)$.

Remark 4.8 Another "standard" regularity theorem shows that

$$D(S) = \{u \in H^2(\Omega) \mid \gamma_1 u = 0\},$$

and the notion of the trace of the normal derivative $u \mapsto \gamma_1 u$ for $u \in H^2(\Omega)$ is more standard; this trace can be shown to be in $H^{1/2}(\partial\Omega)$.

4.5 Notes

Hardy's inequality is named after G. H. Hardy. The original version was that if f is an integrable function with nonnegative values, then

$$\int_0^\infty \left(\frac{1}{x} \int_0^x f(t)\,dt \right)^p dx \le \left(\frac{p}{p-1} \right)^p \int_0^\infty f(x)^p\,dx.$$

Hardy's inequality was first published (without proof) in 1920 in a note by Hardy [Har]. The version presented here can be found, for example, in Kato's book ([Ka], pp. 305–307).

The method of differential quotients mentioned in Section 4.4.2 is described in [LiMa]. For the general properties of the harmonic oscillator, readers can also refer to the book [He1], which presents pseudo-differential techniques.

In the case of Section 4.4, readers who are not familiar with Sobolev spaces and the definition of the trace should start by reading Brézis's book [Br] (see also the publications by Lions and Magenes [LiMa], [Lio1], and [Lio2] or, better, [DaLi, Vol. 4, pp. 1222–1225]) for a detailed explanation.

Regularity theorems for boundary value problems can also be found in [LiMa] and [GiT].

4.6 Exercises

Exercise 4.1 (The free Dirac operator.) Consider on $L^2(\mathbb{R}^2; \mathbb{C}^2)$ the operator $\sum_{j=1}^2 \alpha_j D_{x_j}$ (with $D_{x_j} = -i\partial_{x_j}$) with domain $\mathcal{S}(\mathcal{R}^2; \mathbb{C}^2)$. Here, the α_j are 2×2 Hermitian matrices such that

$$\alpha_j \cdot \alpha_k + \alpha_k \cdot \alpha_j = 2\delta_{jk} I_{\mathbb{C}^2}.$$

Show that this operator is symmetric but not semibounded. This operator is called the Dirac operator. Its domain is $H^1(\mathbb{R}^2, \mathbb{C}^2)$, and its square is the Laplacian:

$$\left(\sum_{j=1}^{2} \alpha_j D_{x_j}\right)^2 = (-\Delta) \otimes I_{\mathbb{C}^2}.$$

Exercise 4.2 (A natural problem in Bose–Einstein theory.) Let $\omega > 0$. Discuss as a function of $\Omega \in \mathbb{R}$ the semiboundedness of the operator defined on $\mathbb{S}(\mathbb{R}^2)$ by

$$H^{\Omega} := -\frac{1}{2}\Delta_{x,y} + \frac{1}{2}\omega^2 r^2 - \Omega L_z, \tag{4.6.1}$$

with

$$L_z = i(x\partial_y - y\partial_x), \tag{4.6.2}$$

The answer can be found by showing that

$$\phi_{j,k}(x, y) = e^{(\omega/2)(x^2+y^2)} (\partial_x + i\partial_y)^j (\partial_x - i\partial_y)^k \left(e^{-\omega(x^2+y^2)}\right), \tag{4.6.3}$$

where j and k are nonnegative integers, is a joint eigenfunction of H^{Ω} and L_z.

5

Compact operators: general properties and examples

We recall briefly the basic properties of compact operators and their spectral theory. We shall emphasize examples.

5.1 Definition and properties

Let us recall the definition of a compact operator.

Definition 5.1 An operator T from a Banach space E into a Banach space F is compact if the range of the unit ball in E by T is relatively compact in F.

We denote by $\mathcal{K}(E, F)$ the space of compact operators, which is a closed subspace in $\mathcal{L}(E, F)$. When $F = E$, we write more briefly $\mathcal{K}(E)$.

There is an alternative, equivalent definition when E and F are Hilbert spaces, which uses sequences.

Proposition 5.2 *If E and F are Hilbert spaces, an operator is compact if and only if, for any sequence x_n that converges weakly in E, $T x_n$ is a strongly convergent sequence in F.*

Here we recall the following definition.

Definition 5.3 Let $\mathcal{H}$ be a Hilbert space. A sequence $(x_n)_{n \in \mathbb{N}}$ is said to be weakly convergent in $\mathcal{H}$ if, for any $y \in \mathcal{H}$, the sequence $\langle x_n, y \rangle_{\mathcal{H}}$ is convergent.

Such a sequence is bounded (Banach–Steinhaus theorem), and we recall that in this case there exists a unique $x \in \mathcal{H}$ such that $\langle x_n, y \rangle_{\mathcal{H}} \to \langle x, y \rangle_{\mathcal{H}}$ for all $y \in \mathcal{H}$. In this case, we write $x_n \rightharpoonup y$.

Let us recall that when one composes a continuous operator and a compact operator (in any order), one obtains a compact operator. This is one possible way to prove compactness.

Another efficient way to prove the compactness of an operator T is to show that it is the limit (for the norm convergence) of a sequence of continuous operators with finite rank (that is, whose range is a finite-dimensional space). We observe that a continuous operator with finite rank is clearly a compact operator (in a finite-dimensional space, closed bounded sets are compact).

5.2 Precompactness

We assume that the reader is aware of the basic results concerning compact sets in metric spaces. In particular, we recall that in a complete metric space E, an efficient way to show that a subset M is relatively compact is to show that for any $\epsilon > 0$, one can cover M with a finite family of balls of radius ϵ in E.

The second standard point is Ascoli's theorem, which gives a criterion for a bounded subset in $C^0(K)$ (where K is compact in $\mathbb{R}^m$) to be relatively compact in terms of uniform equicontinuity. Ascoli's theorem gives, in particular:

- the compact injection of $C^1(K)$ into $C^0(K)$;
- the compact injection of $H^m(\Omega)$ into $C^0(K)$, for $m > n/2$ and $K = \overline{\Omega}$.

Let us recall, finally, a general proposition that permits us to show that a subset in L^2 is relatively compact.

Proposition 5.4 *Let $A \subset L^2(\mathbb{R}^m)$. Assume that:*

1. A is bounded in $L^2(\mathbb{R}^m)$; that is, there exists $M > 0$ such that

$$||u||_{L^2} \leq M, \ \forall u \in A.$$

2. The expression $\epsilon(u, R) := \int_{|x| \geq R} |u(x)|^2 \, dx$ tends to zero as $R \to +\infty$, uniformly with respect to $u \in A$.

3. For $h \in \mathbb{R}^m$, let τ_h be defined on L^2 by $(\tau_h u)(x) = u(x - h)$. Then the expression $\delta(u, h) := ||\tau_h u - u||_{L^2}$ tends to zero as $h \to 0$, uniformly with respect to $u \in A$.

Then A is relatively compact in L^2.

This proposition can be applied to show:

- The compact injection of $H_0^1(\Omega)$ in $L^2(\Omega)$ when Ω is regular and bounded.
- The compact injection of $H^1(\Omega)$ in $L^2(\Omega)$ when Ω is regular and bounded.
- The compact injection of $B^1(\mathbb{R}^m)$ in $L^2(\mathbb{R}^m)$.

5.3 Examples

Let us study three examples. The first is rather academic (but related to the Sturm–Liouville theory), the second comes from the spectral theory of the Dirichlet realization of the Laplacian, and the third comes from quantum mechanics.

5.3.1 Continuous kernels

The first example of this type is the operator T_K associated with the continuous kernel K on $[0, 1] \times [0, 1]$. By this we mean that the operator T_K is defined by

$$E \ni u \mapsto (T_K u)(x) = \int_0^1 K(x, y)u(y)\,dy. \qquad (5.3.1)$$

Here, E could be the Banach space $C^0([0, 1])$ (with the Sup norm) or $L^2(]0, 1[)$.

Proposition 5.5 *If the kernel K is continuous, then the operator T_K is compact from E into E.*

There are two standard proofs of this proposition. The first is based on Ascoli's theorem, giving a criterion relating to the equicontinuity of a subset of continuous functions on compact and relatively compact sets in $C^0([0, 1])$. The other is based on the Stone–Weierstrass theorem, which permits one to recover the operator as the limit of a sequence of finite-rank operators T_{K_n} associated with kernels K_n of the form $K_n(x, y) = \sum_{j=1}^{j_n} f_{j,n}(x)g_{j,n}(y)$.

5.3.2 The inverse of the Dirichlet operator

We return to the operator S, which was introduced in the study of the Dirichlet realization. The following proposition can be shown.

Proposition 5.6 *The operator S^{-1} is compact.*

Proof The operator S^{-1} can be considered as the composition of a continuous operator from L^2 into $V = H_0^1(\Omega)$ and the continuous injection of V into $L^2(\Omega)$. If Ω is relatively compact, we know that we have compact injection from $H^1(\Omega)$ into $L^2(\Omega)$. Hence the injection of V into L^2 is compact, and S^{-1} is compact. For the continuity result, we observe that for all $u \in D(S)$,

$$\|Su\|_{\mathcal{H}}\|u\|_{\mathcal{H}} \geq \langle Su, u \rangle = a(u, u) \geq \alpha\|u\|_V^2 \geq \alpha\|u\|_V\|u\|_{\mathcal{H}}.$$

This gives, for all $u \in D(S)$, the inequality

$$\|Su\|_{\mathcal{H}} \geq \alpha\|u\|_V. \qquad (5.3.2)$$

Using the surjectivity of S, we obtain

$$||S^{-1}||_{\mathcal{L}(\mathcal{H},V)} \leq \frac{1}{\alpha}. \tag{5.3.3}$$

$\square$

Note that $\alpha = 1$ in our example, but this part of the proof is completely general. The only thing that we have used in the framework of Section 3.3 is that the injection of V into $\mathcal{H}$ is compact.

Proposition 5.7 *Under the assumptions of Theorem 3.6 and if the injection of V into $\mathcal{H}$ is compact, then S^{-1} is compact from $\mathcal{H}$ into $\mathcal{H}$.*

5.3.3 The inverse of the harmonic oscillator

The analysis is analogous. We have seen that the Sobolev space $H_0^1(\mathbb{R})$ has to be replaced by the space

$$B^1(\mathbb{R}) := \{u \in L^2(\mathbb{R}), \, xu \in L^2 \text{ and } u' \in L^2\}.$$

We can then prove, using a standard precompactness criterion (see Proposition 5.4), that $B^1(\mathbb{R})$ has compact injection in $L^2(\mathbb{R})$. In particular, we have to use the inequality

$$\int_{|x| \geq R} |u(x)|^2 \, dx \leq \frac{1}{R^2} ||u||_{B^1(\mathbb{R})}^2. \tag{5.3.4}$$

It is very important to realize that the space $H^1(\mathbb{R})$ is not compactly injected in L^2. To understand this point, it is enough to consider the sequence $u_n = \chi(x - n)$, where χ is a function in $C_0^\infty(\mathbb{R})$ with norm in L^2 equal to 1. This is a bounded sequence in H^1, which converges weakly in H^1 to 0 and is not convergent in $L^2(\mathbb{R})$.

5.4 Hilbertian adjoints

We recall (see Section 2.2) that the adjoint of a bounded operator in the Hilbertian case is bounded. When E and F are possibly different Hilbert spaces, the Hilbertian adjoint is defined through the identity

$$\langle Tx, y \rangle_F = \langle x, T^*y \rangle_E, \, \forall x \in E, \, \forall y \in F. \tag{5.4.1}$$

Example 5.8 Let Ω be an open set in $\mathbb{R}^m$ and let Π_Ω be the operator of restriction to Ω: $L^2(\mathbb{R}^m) \ni u \mapsto u_{/\Omega}$. Then Π_Ω^* is the operator of extension by 0 outside of Ω.

In a Hilbert space, we have

$$M^\perp = \overline{M}^\perp.$$

In particular, if M is a closed subspace, we have already used the property

$$(M^\perp)^\perp = M. \qquad (5.4.2)$$

In the case of bounded operators ($T \in \mathcal{L}(E, F)$), one can easily obtain the properties

$$N(T^*) = R(T)^\perp \qquad (5.4.3)$$

and

$$\overline{R(T)} = (N(T^*))^\perp. \qquad (5.4.4)$$

Let us also recall the following proposition.

Proposition 5.9 *The adjoint of a compact operator is compact.*

5.5 Notes

For the general theory of compact operators, readers are referred to the books by Brézis [Br, chapter VI], Reed and Simon [RS-I], and Lévy-Bruhl [Le-Br]. The proof of Proposition 5.4 can be found, for example, in [Br, section IV.5].

5.6 Exercises

Exercise 5.1 Using the theory of differential equations, recover the result of Section 5.3.2 by computing the distribution kernel of S^{-1}, where S is the Dirichlet realization of $-d^2/dx^2 + 1$ on $]0, 1[$.

Exercise 5.2 Let $(V(n))_n$ be a sequence indexed by $\mathbb{N}$. Let S_V be the operator of multiplication operator by V on $\ell^2(\mathbb{N})$, where $\ell^2(\mathbb{N})$ is the space of the sequences $(u_n)_{n \in \mathbb{N}}$ such that $\sum_{n \in \mathbb{N}} |u_n|^2 < +\infty$. Show that if $\lim_{n \to +\infty} V(n) = 0$, then S_V is a compact operator in $K(\ell^2(\mathbb{N}))$.

Exercise 5.3 Show that the operator T, defined on $L^2(]0, 1[)$ by $Tf(x) = \int_0^x f(t)\,dt$, is a compact operator. Determine its adjoint. Is T compact on $L^2(\mathbb{R}^+)$?

Exercise 5.4 Let $\mathbf{K_f}$ be the operator defined on $L^2(\mathbb{S}^1)$ as the convolution with a function f in L^1. Show that $\mathbf{K_f}$ is a compact operator.

Exercise 5.5 Let γ_0 be the trace operator on $x_m = 0$ defined from $H^1(\mathbb{R}^m_+)$ onto $H^{1/2}(\mathbb{R}^{m-1})$. Determine the adjoint.

Solution of Exercise 5.5

We treat the case $m = 1$ here. $u \mapsto u(0)$ is a continuous linear form on $H^1(\mathbb{R}^+)$. We can, for example, rewrite it using the formula $u(0) = -\int_0^{+\infty} (\chi u)'(t)\,dt$, where χ is a C^∞ function with support in, say, $[0,1]$ such that $\chi(0) = 1$. By Riesz's representation theorem, there exists $f \in H^1(\mathbb{R}^+)$ such that

$$u(0) = \int_0^{+\infty} u'(t)\bar{f}'(t)\,dt + \int_0^{+\infty} u(t)\bar{f}(t)\,dt, \ \forall u \in H^1(\mathbb{R}^+).$$

It can be verified that $f(t) = \exp{-t}$. The adjoint of γ_0 is $\alpha \mapsto \alpha \exp{-t}$.

Let us also give a hint for the general case. For $v \in H^{1/2}$ and $u \in H^1$, we write

$$\langle u(0,\cdot), v(\cdot)\rangle = \int\int_0^{+\infty} \langle \xi'\rangle^2 f(t\langle \xi'\rangle)\,\hat{u}(\xi',t)\,\overline{\hat{v}(\xi')}\,d\xi'\,dt$$

$$+ \int\int_0^{+\infty} \langle \xi'\rangle\,f'(t\langle \xi'\rangle)\,\partial_t\hat{u}(\xi',t)\,\overline{\hat{v}}\,d\xi'\,dt,$$

with $\langle \xi'\rangle = \sqrt{1 + |\xi'|^2}$. The adjoint of γ_0 is then defined by

$$v \mapsto (\gamma_0^* v)(x', x_n) = \int e^{ix.'\xi'}\,e^{-x_n\langle \xi'\rangle}\,\hat{v}(\xi')\,d\xi'.$$

6

Spectral theory for bounded operators

If the spectral theory for linear operators in $\mathcal{L}(\mathcal{H})$ is easy in the case where $\mathcal{H}$ is finite dimensional, the infinite-dimensional case gives rise to many different situations, which will be described in this chapter.

6.1 Fredholm's alternative

Let us first recall Riesz's theorem.

Theorem 6.1 *Let E be a normed linear space such that $\overline{B_E}$ is compact, then E is finite dimensional.*

Let us now describe Fredholm's alternative.

Theorem 6.2 *Let $T \in \mathcal{K}(E)$. Then:*

1. *$N(I - T)$ is finite dimensional.*
2. *$R(I - T)$ is closed (of finite codimension).*
3. *$R(I - T) = E$ if and only if $N(I - T) = \{0\}$.*

We shall use this theorem only in the Hilbertian framework, so $E = \mathcal{H}$, and, for simplicity, we shall prove it under the additional assumption that $T = T^*$.

Proof We divide the proof into successive steps.

Step 1. Part 1 of the theorem is a consequence of Riesz's theorem applied to $E = N(I - T)$.

Let $x_n \in B_E$. Since T is a compact operator, we can extract a subsequence x_{n_j} such that $T x_{n(j)}$ is convergent. But $x_{n(j)} = T x_{n(j)}$, and hence $x_{n(j)}$ is convergent. This shows that $\overline{B_E}$ is compact.

Step 2. We show that $R(I - T)$ is closed.

Let y_n be a sequence in $R(I - T)$ such that $y_n \to y$ in $\mathcal{H}$. We would like to show that $y \in R(I - T)$. Let x_n be such that $y_n = x_n - Tx_n$. Of course, there is no restriction in assuming in addition that $x_n \in N(I - T)^{\perp}$.

Step 2a. Let us first show the weaker property that the sequence x_n is bounded.

Suppose that there exists a subsequence x_{n_j} such that $\|x_{n_j}\| \to +\infty$. Considering $u_{n_j} = x_{n_j}/\|x_{n_j}\|$, we observe that

$$(*) \quad u_{n_j} - Tu_{n_j} \to 0.$$

Since the sequence is bounded, we observe that (after possibly extracting a subsequence) we can consider the sequence u_{n_j} to be weakly convergent. This implies that Tu_{n_j} is convergent (T is compact). Now, using $(*)$, we obtain the convergence of u_{n_j} to u:

$$u_{n_j} \to u, \ Tu = u, \ \|u\| = 1.$$

But $u \in N(I - T)^{\perp}$, and hence we obtain $u = 0$ and a contradiction.

Step 2b. We have consequently obtained the result that the sequence x_n is bounded. We can therefore extract a subsequence x_{n_j} that converges weakly to x_{∞} in $\mathcal{H}$. Using the compactness of T, we obtain the result that Tx_{n_j} converges strongly to Tx_{∞}. Hence the sequence x_{n_j} tends strongly to $y + Tx_{\infty}$.

We have, finally,

$$y + Tx_{\infty} = x_{\infty},$$

and consequently have proved that $y = x_{\infty} - Tx_{\infty}$.

Step 3. If $N(I - T) = \{0\}$, then $N(I - T^*) = \{0\}$ (here we use our additional assumption for simplification) and, since $R(I - T)$ is closed, we obtain

$$R(I - T) = N(I - T^*)^{\perp} = \mathcal{H}.$$

The converse can also be seen immediately, as $T = T^*$.

Step 4. We have

$$R(I - T)^{\perp} = N(I - T^*) = N(I - T).$$

This shows, according to Part 1, that $R(I - T)$ is of finite codimension (second statement of Part 2).

This ends the proof of Fredholm's alternative in the particular case where T is self-adjoint. $\square$

Remark 6.3 Under the same assumptions, it is possible to show that

$$\dim N(I - T) = \dim N(I - T^*).$$

6.2 Index theory

We now present another approach, which permits us to develop index theory. Let E and F be two Hilbert spaces and let $T \in \mathcal{L}(E, F)$. We say that T is a semi-Fredholm operator if the kernel of T, $N(T)$, is finite dimensional and if $R(T)$ is closed. We say that T is a Fredholm operator if it is a semi-Fredholm operator and $N(T^*)$ is finite dimensional. We have shown in Theorem 6.2 that if K is a compact operator, $I + K$ is a Fredholm operator. We shall give an independent proof of this result in this section.

We define the index of T by

$$\mathrm{Ind}\,(T) = \dim N(T) - \dim N(T^*).$$

This is an integer if T is a Fredholm operator. When the operator T is semi-Fredholm and not Fredholm, we define the index by $\mathrm{Ind}\,(T) = -\infty$.

Proposition 6.4 *Let* $T \in \mathcal{L}(E, F)$. *Then the following statements are equivalent:*

(P1) The operator T is semi-Fredholm.

(P2) From any bounded sequence of E whose range with respect to T is convergent, one can extract a convergent subsequence.

Proof

(P2) implies (P1).

Suppose that (P2) is satisfied. The first point is to observe that $N(T)$ is finite dimensional. Its unit ball is evidently relatively compact, by (P2), and Riesz's theorem applies.

Now let E_0 be the orthogonal of $N(T)$ in E. We observe that (P2) also implies the existence of $C > 0$ such that

$$\|u\| \leq C\,\|Tu\|, \ \forall u \in E_0 \tag{6.2.1}$$

(in contradiction). We then note that since the range $R(T)$ of T and that of $T_{/E_0}$ are equal, $R(T)$ is closed (by an argument[1] applied to $T_{/E_0}$).

---

[1] If E is a Banach space, $A \in \mathcal{L}(E)$, then A is injective **and** with a closed range if and only if there exists $C > 0$ such that $\|Tu\| \geq C\|u\|$, for all $u \in E$.

(P1) implies (P2).

Conversely, (P1) implies (6.2.1) (by a previous argument). Now consider a bounded sequence $(u_n)_{n \in \mathbb{N}}$ such that $(T u_n)$ is convergent. Then we can write u_n as the sum

$$u_n = v_n + w_n,$$

with v_n in $N(T)$ and w_n in E_0. It is clear, by (6.2.1), that w_n is a Cauchy sequence, and since $N(T)$ is finite dimensional, we can extract a subsequence such that v_{n_k} is convergent (we note here that $||v_n|| \leq ||u_n||$, so v_n is a bounded sequence). The sequence $u_{n_k} = v_{n_k} + w_{n_k}$ answers the question. $\qquad \square$

Proposition 6.5 *The vector space* $\mathcal{F}red(E, F)$ *of the Fredholm operators from E into F is open in $\mathcal{L}(E, F)$. The index is constant in each connected component of $\mathcal{F}red(E, F)$.*

Proof Let $T \in \mathcal{F}red(E, F)$. We introduce the operator $\mathcal{T}$ in $\mathcal{L}(E \times N(T^*); F \times N(T))$, which is defined by

$$(u, x) \mapsto (T u + R_+ x, R_- u),$$

where R_+ is the canonical injection of $N(T^*)$ in F and R_- is the orthogonal projection from E onto $N(T)$. It is easy to verify that $\mathcal{T}$ is effectively in $\mathcal{L}(E \times N(T^*); F \times N(T))$ and bijective. By Banach's theorem, it admits a continuous inverse $\mathcal{E}_0$ in $\mathcal{L}(F \times N(T); E \times N(T^*))$, which we can rewrite in the form

$$\mathcal{E}_0 := \begin{pmatrix} E_0 & E_0^+ \\ E_0^- & 0 \end{pmatrix}.$$

Remark 6.6 If we write $\mathcal{E}_0 \circ \mathcal{T} = I$, we obtain in particular the fact that if T is a Fredholm operator, then there exists a bounded operator E_0 and a compact operator K (with finite rank) such that

$$E_0 T = I + K.$$

Similarly, there exists a bounded operator E_0 and a compact operator K' (with finite rank) such that

$$T E_0 = I + K'.$$

Using Proposition 6.4, we can show that the converse is true.

We now return to the analysis of $(T + S)$ for $S \in \mathcal{L}(E, F)$.

We introduce

$$(\mathcal{T}_S)(u, x) = ((T + S)u + R_+ x, R_- u).$$

We can easily see that there exists $\epsilon > 0$ such that if $||S|| \leq \epsilon$, then $\mathcal{T}_S$ is invertible. Moreover, its inverse $\mathcal{E}_S$ is continuous and can be written in the block form

$$\mathcal{E}_S := \begin{pmatrix} E_S & E_S^+ \\ E_S^- & E_S^{+-} \end{pmatrix}.$$

We now write that $\mathcal{E}_S$ is the left inverse and the right inverse of $\mathcal{T}_S$. We obtain

$$\begin{array}{ll}
E_S(T + S) + E_S^+ R_- = I, & \text{(a)} \\
E_S^-(T + S) + E_S^{+-} R_- = 0, & \text{(b)} \\
E_S R_+ = 0, & \text{(c)} \\
E_S^- R_+ = I, & \text{(d)}
\end{array} \qquad (6.2.2)$$

and

$$\begin{array}{ll}
(T + S)E_S + R_+ E_S^- = I, & \text{(a)} \\
(T + S)E_S^+ + R_+ E_S^{+-} = 0, & \text{(b)} \\
R_- E_S^+ = I, & \text{(c)} \\
R_- E_S = 0. & \text{(d)}
\end{array} \qquad (6.2.3)$$

Let us show that $N(T+S)$ is finite dimensional. If $u \in N(T+S)$, (6.2.2)(b) shows that $R_- u \in N(E_S^{+-})$ and (6.2.2)(a) shows that R_- is injective on $N(T + S)$. So, we obtain

$$\dim N(T + S) \leq \dim N(E_S^{+-}) \leq \dim N(T). \qquad (6.2.4)$$

This shows in particular that $N(T + S)$ is finite dimensional.

Let us now show that $(T + S)$ has a closed range. We use the criterion given in Proposition 6.4. Let u_n be a bounded sequence such that $(T + S)u_n$ is convergent. Then (6.2.2)(a) implies

$$u_n = E_S(T + S)u_n - R_- u_n := v_n - w_n.$$

But $v_n = E_S(T + S)u_n$ is convergent, and $w_n = R_- u_n$ is a bounded sequence in $N(T)$ that is finite dimensional. We can then extract a convergent subsequence.

If we observe that $\mathcal{E}_S^*$ is the inverse of $\mathcal{T}_S^*$, we immediately obtain the result that $(T + S)^*$ has a closed range and that

$$\dim N\left((T + S)^*\right) \leq \dim N((E_S^{+-})^*) \leq \dim N(T^*). \qquad (6.2.5)$$

In particular, dim $N((T+S)^*)$ is finite and $(T+S)^*$ is Fredholm. (On the way, we have proved that if T is Fredholm, then T^* is Fredholm.)

It remains to compute the index.

Let $y \in N(E_S^{+-})$. Using (6.2.3)(b), we deduce that $E_S^+ y \in N(T+S)$. Moreover, (6.2.3)(c) implies that E_S^+ is injective. Hence we obtain

$$\dim N(E_S^{+-}) \leq \dim N(T+S), \qquad (6.2.6)$$

and, bearing in mind (6.2.4),

$$\dim N(E_S^{+-}) = \dim N(T+S) \leq \dim N(T).$$

Working with the adjoint, we also obtain

$$\dim N((E_S^{+-})^*) = \dim N((T+S)^*) \leq \dim N(T^*). \qquad (6.2.7)$$

This implies the following interesting property:

$$\text{Ind}\,(T+S) = \text{Ind}\,E_S^{+-}. \qquad (6.2.8)$$

Hence it remains to compute the index of $E_S^{+-} \in \mathcal{L}(N(T), N(T^*))$, which is standard in finite dimension. We have in fact

$$\text{Ind}\,(E_S^{+-}) = \dim N(T) - \dim N(T^*).$$

This is almost the end of the proof of the proposition. We have proved the existence of $\epsilon > 0$ such that if $\|S\| < \epsilon$, then $\text{Ind}\,(T+S) = \text{Ind}\,(T)$. Hence the map $T \mapsto \text{Ind}\,(T)$ from $\mathcal{F}red(E, F)$ into $\mathbb{Z}$ is locally constant.

The next theorem is just a small extension of Theorem 6.2.

Theorem 6.7 *Let K be a compact operator; then $I + K$ is Fredholm and its index is 0.*

We use Proposition 6.4 for $I + K$ and $I + K^*$ for the first assertion. For the second assertion, we introduce the family of operators $[0, 1] \ni t \mapsto I + tK$. We note that the index is 0 for $t = 0$ because the index of the identity equals 0. On the other hand, the index is locally constant, and hence constant in each connected component. Consequently, it equals 0 in the component of the identity.

It follows from Theorem 6.7 that $N(I + K) = 0$ if and only if $N(I + K^*) = 0$.

Proposition 6.8 *Let* $T \in \mathcal{F}red(E, F)$ *and* $S \in \mathcal{F}red(F, G)$. *Then* $S \circ T \in \mathcal{F}red(E, G)$ *and* $\mathrm{Ind}(S \circ T) = \mathrm{Ind}\, S + \mathrm{Ind}\, T$.

Proof We apply Proposition 6.4. Let x_n be a bounded sequence such that $ST x_n$ is convergent. As S is Fredholm and $T x_n$ is bounded, there exists a convergent subsequence such that $T x_{n_k}$ is convergent. We apply the same argument with T and the sequence x_{n_k}.

　　To compute the index, we consider the family of operators in $\mathcal{L}(E \times F; F \times G)$

$$\left[0, \frac{\pi}{2} \right] \ni t \mapsto \mathcal{P}_t := \begin{pmatrix} I_F & 0 \\ 0 & S \end{pmatrix} \begin{pmatrix} \cos t \, I_F & \sin t \, I_F \\ -\sin t \, I_F & \cos t \, I_F \end{pmatrix} \begin{pmatrix} T & 0 \\ 0 & I_F \end{pmatrix}.$$

We observe that for $t = 0$, we have

$$\mathcal{P}_0 = \begin{pmatrix} T & 0 \\ 0 & S \end{pmatrix},$$

whose index is $\mathrm{Ind}\, T + \mathrm{Ind}\, S$, and that for $t = \pi/2$ we have

$$\mathcal{P}_{\pi/2} = \begin{pmatrix} 0 & I_F \\ -ST & 0 \end{pmatrix}.$$

These two operators are in the same component and hence have the same index, and this index is clearly $\mathrm{Ind}\,(ST)$. □

□

6.3 Resolvent set for bounded operators

In this section, E could be a Banach space on $\mathbb{R}$ or $\mathbb{C}$, but we need only the Hilbertian case in the applications treated in this book.

Definition 6.9 (Resolvent set.) For $T \in \mathcal{L}(E)$, the resolvent set is defined by

$$\varrho(T) = \{\lambda \in \mathbb{C}; \ (T - \lambda I) \text{ is bijective from } E \text{ on } E\}. \tag{6.3.1}$$

　　Note that in this case $(T - \lambda I)^{-1}$ is continuous (Banach's theorem). It is easy to see that $\varrho(T)$ is an open set in $\mathbb{C}$. If $\lambda_0 \in \varrho(T)$, we observe that

$$(T - \lambda) = (T - \lambda_0) \left(Id + (\lambda - \lambda_0)(T - \lambda_0)^{-1} \right). \tag{6.3.2}$$

Hence $(T - \lambda)$ is invertible if $|\lambda - \lambda_0| < ||(T - \lambda_0)^{-1}||^{-1}$. We also obtain the following identity for all λ, $\lambda_0 \in \varrho(T)$:

$$(T - \lambda)^{-1} - (T - \lambda_0)^{-1} = (\lambda - \lambda_0)(T - \lambda)^{-1}(T - \lambda_0)^{-1}. \qquad (6.3.3)$$

Note also that $\varrho(T)$ cannot be $\mathbb{C}$. If $\varrho(T)$ were $\mathbb{C}$, then $z \mapsto (z - T)^{-1}$ would be a bounded holomorphic function on $\mathbb{C}$ with values in $\mathcal{L}(E)$.

Definition 6.10 (Spectrum.) The spectrum of T, $\sigma(T)$, is the complementary set of $\varrho(T)$ in $\mathbb{C}$.

Note that $\sigma(T)$ is a closed set in $\mathbb{C}$. This is typically the case when T is a compact injective operator in a Banach space of infinite dimension.

We say that λ is an eigenvalue if $N(T - \lambda I) \neq 0$. $N(T - \lambda I)$ is called the eigenspace associated with λ.

Definition 6.11 (Point spectrum.) The point spectrum $\sigma_p(T)$ of T is defined as the set of the eigenvalues of T.

Example 6.12 (Basic example.) Let $\mathcal{H} = L^2(]0, 1[)$ and $f \in C^0([0, 1])$. Let T_f be the operator of multiplication by f. Then one has

$$\sigma(T_f) = \text{Im } f =: \{\lambda \in \mathbb{C} \mid \exists x \in [0, 1] \text{ with } f(x) = \lambda\}.$$

$$\sigma_p(T_f) = \text{Sta}(f) =: \{\lambda \in \mathbb{C} \mid \text{meas}(f^{-1}(\lambda)) > 0\}.$$

For the first assertion, we can immediately see that if $\lambda \notin \text{Im } f$, then $T_{(f-\lambda)^{-1}}$ is a continuous inverse of $T_f - \lambda$. On the other hand, if $\lambda = f(x_0)$ for some $x_0 \in]0, 1[$, then we have $(T_f - \lambda)u_n \to 0$ and $||u_n|| = 1$ for $u_n = (1/\sqrt{n})\chi((x - x_0)/n)$, where χ is a C_0^∞ function such that $||\chi|| = 1$. This shows that $f(]0, 1[) \subset \sigma(T_f)$, and we can conclude by considering the closure.

Note that the point spectrum is not necessarily a closed set. Note also that one can have a strict inclusion of the point spectrum in the spectrum, as can be observed in the previous example (when f is strictly monotonic) and in the following example.

Example 6.13 Let $E = \ell^2(\mathbb{N})$, and let τ be the shift operator defined by

$$(\tau u)_0 = 0, \quad (\tau u)_n = u_{n-1}, \, n > 0,$$

where $u = (u_0, \cdots, u_n, \cdots) \in \ell^2(\mathbb{N})$. Then it is easy to see that τ is injective. Hence 0 is not an eigenvalue and not surjective $((1, 0, \cdots)$ is not in the range of $\tau)$, and hence 0 is in the spectrum of τ.

Another interesting example is the following.

Example 6.14 Let $E = \ell^2(\mathbb{Z}, \mathbb{C})$, and let T be the operator defined by

$$(Tu)_n = \frac{1}{2} (u_{n-1} + u_{n+1}), \, n \in \mathbb{Z},$$

for $u = (u_n)_{n \in \mathbb{Z}} \in \ell^2(\mathbb{Z})$. Using the isomorphism between $\ell^2(\mathbb{Z}; \mathbb{C})$ and $L^2(S^1; \mathbb{C})$, which associates with the sequence $(u_n)_{n \in \mathbb{Z}}$ the function $\sum_{n \in \mathbb{Z}} u_n \exp in\theta$, we have to analyze the operator of multiplication by $\cos \theta$,

$$f \mapsto \mathcal{T}f = \cos \theta f.$$

The problem is then analyzed as in Example 6.12. We conclude that the spectrum of $\mathcal{T}$ is $[-1, +1]$, that T has no eigenvalues, and that its spectrum is $[-1, +1]$.

Proposition 6.15 *The spectrum* $\sigma(T)$ *is a compact set included in the ball* $\overline{B}(0, \|T\|)$.

This proposition follows immediately if we observe that $(I - T/\lambda)$ is invertible if $|\lambda| > \|T\|$.

6.4 Spectral theory for compact operators

In the case of a compact operator, we have a more precise description of the spectrum.

Theorem 6.16 *Let* $T \in \mathcal{K}(E)$*, where* E *is an infinite-dimensional Banach space. Then*

1. $0 \in \sigma(T)$.

2. $\sigma(T) \setminus \{0\} = \sigma_p(T) \setminus \{0\}$.

3. We are in one (and only one) of the following cases:

  - $\sigma(T) = \{0\}$;
  - $\sigma(T) \setminus \{0\}$ *is finite;*

- *or $\sigma(T) \setminus \{0\}$ can be described as a sequence of distinct points tending to 0.*

4. *Each $\lambda \in \sigma_p(T) \setminus \{0\}$ is isolated, and* dim $N(T - \lambda I) < +\infty$.

Proof

(a) If $0 \notin \sigma(T)$, then T admits a continuous inverse T^{-1}, and the identity, which can be considered as

$$I = T \circ T^{-1},$$

is a compact operator, as it is the composition of the compact operator T and the continuous operator T^{-1}. Using Riesz's theorem, we obtain a contradiction in the case where E is supposed to be of infinite dimension.

(b) The fact that if $\lambda \neq 0$ and $\lambda \in \sigma(T)$, then λ is an eigenvalue comes directly from Fredholm's alternative applied to $(I - T/\lambda)$.

(c) The last step comes essentially from the following lemma.

Lemma 6.17 *Let $(\lambda_n)_{n \geq 1}$ be a sequence of distinct points in $\sigma(T) \setminus \{0\}$, such that $\lambda_n \to \lambda$. Then $\lambda = 0$.*

Proof We shall just give the proof in the Hilbertian case. For all $n > 0$, let e_n be a normalized eigenvector such that $(T - \lambda_n)e_n = 0$, and let E_n be the vector space spanned by $\{e_1, e_2, \cdots, e_n\}$. Let us show that we have a strict inclusion of E_n in E_{n+1}. We prove this point by recursion.

Let us assume the result up to order $n - 1$ and show it at order n. If, in contradiction, $E_{n+1} = E_n$, then $e_{n+1} \in E_n$ and we can write

$$e_{n+1} = \sum_{j=1}^{n} \alpha_j e_j.$$

We apply T to this relation. Using the property that e_{n+1} is an eigenvector, we obtain

$$\lambda_{n+1} \left(\sum_{j=1}^{n} \alpha_j e_j \right) = \sum_{j=1}^{n} \alpha_j \lambda_j e_j.$$

Using the recursion assumption, $\{e_1, \cdots, e_n\}$ is a basis of E_n, and since the λ_j are distinct, we obtain $\alpha_j = 0$ for all $j = 1, \cdots, n$ and a contradiction with the normalization $\|e_{n+1}\| = 1$.

So, we can find a sequence u_n such that $u_n \in E_n \cap E_{n-1}^{\perp}$ and $\|u_n\| = 1$. Since T is compact, we can extract a convergent subsequence (still denoted by Tu_n) from the sequence (Tu_n). If $\lambda_n \to \lambda \neq 0$, we would also have the convergence of this subsequence $((1/\lambda_n)Tu_n)$ and consequently the Cauchy property.

Let us show that this leads to a contradiction. We note that

$$(T - \lambda_n)E_n \subset E_{n-1}.$$

Let $n > m \geq 2$. We have

$$||\frac{Tu_n}{\lambda_n} - \frac{Tu_m}{\lambda_m}||^2 = ||\frac{(T-\lambda_n)u_n}{\lambda_n} - \frac{(T-\lambda_m)u_m}{\lambda_m} + u_n - u_m||^2 \quad (6.4.1)$$

$$= ||\frac{(T-\lambda_n)u_n}{\lambda_n} - \frac{(T-\lambda_m)u_m}{\lambda_m} - u_m||^2 + ||u_n||^2 \quad (6.4.2)$$

$$\geq ||u_n||^2 = 1. \quad (6.4.3)$$

Consequently, we cannot extract a Cauchy subsequence from the sequence $(1/\lambda_n)Tu_n$. This contradicts the assumption $\lambda \neq 0$.

This ends the proof of the lemma and of the theorem. □

□

We shall now consider the Hilbertian case and see what new properties can be obtained by using the additional assumption that T is self-adjoint.

6.5 Spectrum of a self-adjoint operator

As $T = T^*$, the spectrum is real. If $\operatorname{Im} \lambda \neq 0$, we can immediately verify that

$$|\operatorname{Im} \lambda| \, ||u||^2 \leq |\operatorname{Im} \langle (T - \lambda)u, u \rangle| \leq ||(T - \lambda)u|| \cdot ||u||, \forall u \in \mathcal{H}.$$

This implies that the map $(T - \lambda)$ is injective and has a closed range. But the orthogonal of the range of $(T - \lambda)$ is the kernel of $(T - \bar{\lambda})$, which is reduced to 0. So $(T - \lambda)$ is bijective.

This is actually a consequence of the Lax–Milgram theorem (Theorem 3.6) (in the simple case where $V = \mathcal{H}$), once we have observed the inequality

$$|\langle (T - \lambda)u, u \rangle| \geq |\operatorname{Im} \lambda| \, ||u||^2.$$

Using the Lax–Milgram theorem again, we can show the following.

Theorem 6.18 *Let* $T \in \mathcal{L}(\mathcal{H})$ *be a self-adjoint operator. Then the spectrum of* T *is contained in* $[m, M]$, *with* $m = \inf\langle Tu, u\rangle / \|u\|^2$ *and* $M = \sup\langle Tu, u\rangle / \|u\|^2$. *Moreover,* m *and* M *belong to the spectrum of* T.

Proof We have already mentioned that the spectrum is real. If λ is real and $\lambda > M$, we can then apply the Lax–Milgram theorem to the sesquilinear form $(u, v) \mapsto \lambda\langle u, v\rangle - \langle Tu, v\rangle$.

Let us now show that $M \in \sigma(T)$.

We observe that, by the Cauchy–Schwarz inequality applied to the scalar product $(u, v) \mapsto M\langle u, v\rangle - \langle Tu, v\rangle$, we have

$$|\langle Mu - Tu, v\rangle| \leq \langle Mu - Tu, u\rangle^{1/2} \langle Mv - Tv, v\rangle^{1/2}.$$

In particular, we obtain

$$\|Mu - Tu\|_{\mathcal{H}} \leq \|M - T\|_{\mathcal{L}(\mathcal{H})}^{1/2}\langle Mu - Tu, u\rangle^{1/2}. \tag{6.5.1}$$

Let $(u_n)_{n \in \mathbb{Z}}$ be a sequence such that $\|u_n\| = 1$ and $\langle Tu_n, u_n\rangle \to M$ as $n \to +\infty$. By (6.5.1), we obtain the result that $(T - M)u_n$ tends to 0 as $n \to +\infty$. This implies that $M \in \sigma(T)$. If this were not the case, we would obtain the result that $u_n = (M - T)^{-1}((M - T)u_n)$ tends to 0, in contradiction with $\|u_n\| = 1$. $\qquad\square$

This theorem has the following important corollary.

Corollary 6.19 *Let* $T \in \mathcal{L}(\mathcal{H})$ *be a self-adjoint operator such that* $\sigma(T) = \{0\}$. *Then* $T = 0$.

We can show first that $m = M = 0$ and, consequently, that $\langle Tu, u\rangle_{\mathcal{H}} = 0$ for all $u \in \mathcal{H}$. But $\langle Tu, v\rangle$ can be written as a linear combination of terms of the type $\langle Tw, w\rangle_{\mathcal{H}}$, and this gives the result by taking $v = Tu$.

Another related property that is useful is the following.

Proposition 6.20 *If* T *is nonnegative and self-adjoint, then* $\|T\| = M$.

The proof is quite similar. We observe (by the Cauchy–Schwarz inequality for the sesquilinear form $(u, v) \mapsto \langle Tu, v\rangle$) that

$$|\langle Tu, v\rangle| \leq \langle Tu, u\rangle^{1/2} \langle Tv, v\rangle^{1/2}.$$

This implies, using Riesz's theorem,

$$\|Tu\| \leq \|T\|^{1/2}(\langle Tu, u\rangle)^{1/2}.$$

Returning to the definition of $||T||$, we obtain

$$||T|| \leq ||T||^{1/2} M^{1/2},$$

and the inequality

$$||T|| \leq M.$$

But it follows immediately that

$$\langle Tu, u \rangle \leq ||T|| \, ||u||^2.$$

This gives the converse inequality and completes the proof of the proposition.

6.6 Spectral theory for compact self-adjoint operators

We have a very precise description of self-adjoint compact operators.

Theorem 6.21 *Let $\mathcal{H}$ be a separable Hilbert space and T a compact self-adjoint operator. Then $\mathcal{H}$ admits a Hilbertian basis consisting of eigenfunctions of T.*

Proof Let $(\lambda_n)_{n \geq 1}$ be the sequence of distinct eigenvalues of T, except 0. Their existence is a consequence of Theorem 6.16, and we also observe that the eigenvalues are real.

We define λ_0 as equal to 0, and define $E_0 = N(T)$ and $E_n = N(T - \lambda_n I)$. We know (from Riesz's theorem) that

$$0 < \dim E_n < +\infty.$$

Let us show that $\mathcal{H}$ is the Hilbertian sum of the $(E_n)_{n \geq 0}$.

(a) The spaces (E_n) are mutually orthogonal. If $u \in E_m$ and $v \in E_n$ with $m \neq n$, we have

$$\langle Tu, v \rangle_{\mathcal{H}} = \lambda_m \langle u, v \rangle_{\mathcal{H}} = \langle u, Tv \rangle_{\mathcal{H}} = \lambda_n \langle u, v \rangle_{\mathcal{H}},$$

and, consequently, $\langle u, v \rangle_{\mathcal{H}} = 0$.

(b) Let F be the vector space spanned by the $(E_n)_{n \geq 0}$. Let us verify that F is dense in $\mathcal{H}$. It is clear that $TF \subset F$ and, using the self-adjoint character of T, we have also $TF^{\perp} \subset F^{\perp}$. The operator $\tilde{T}$ obtained by restriction of T to $F^{\perp}$ is a compact self-adjoint operator. But it can be shown easily that $\sigma(\tilde{T}) = \{0\}$ and, consequently, $\tilde{T} = 0$. But $F^\perp \subset N(T) \subset F$, and hence $F^\perp = \{0\}$. Consequently, F is dense in $\mathcal{H}$.

(c) To achieve the proof, we choose a Hilbertian basis in each E_n. Taking the union of these bases, we obtain a Hilbertian basis of $\mathcal{H}$, effectively formed from eigenfunctions of T. $\square$

Remark 6.22 If T is a compact self-adjoint operator, we can write any u in the form

$$u = \sum_{n=0}^{+\infty} u_n, \quad \text{with } u_n \in E_n.$$

This permits us to write

$$Tu = \sum_{n=1}^{+\infty} \lambda_n u_n.$$

If, for $k \in \mathbb{N}^*$, we define T_k by

$$T_k u = \sum_{n=1}^{k} \lambda_n u_n,$$

we can easily see that T_k is of finite rank and that

$$\|T - T_k\| \leq \sup_{n \geq k+1} |\lambda_n|.$$

Hence the operator T appears as the limit in $\mathcal{L}(\mathcal{H})$ of the sequence T_k, as $k \to +\infty$.

6.7 Canonical form for a non-self-adjoint compact operator

In the non-self-adjoint case, one cannot in general find an orthonormal basis of eigenfunctions, but we have the following substitute.

Proposition 6.23 *Let T be a compact operator in a separable Hilbert space $\mathcal{H}$. Then there exist two orthonormal families $(\varphi_n)_n$ and $(\psi_n)_n$ (the family being finite if and only if T has finite rank) and a real sequence $s_n > 0$ such that T takes the following form in $\mathcal{L}(\mathcal{H})$:*

$$T = \sum_{n} s_n \langle \cdot, \varphi_n \rangle \, \psi_n. \tag{6.7.1}$$

For the proof, we apply the spectral decomposition theorem (Theorem 6.21) to T^*T. Let (φ_n) be the orthonormal system obtained by keeping only the eigenfunctions corresponding to the nonzero eigenvalues, which are denoted by s_n^2 (with $s_n > 0$). This system is an orthonormal basis of $N(T)^\perp$ because $N(T) = N(T^*T)$.

We now set

$$\psi_n = \frac{1}{s_n} T\varphi_n.$$

We can immediately see that (ψ_n) is an orthonormal basis of $\overline{\operatorname{Im} T} = (N(T^*))^\perp$.

It remains to show (6.7.1). It is enough to verify it on each nonzero eigenspace of T^*T. We have, for any ℓ,

$$\sum_n s_n \langle \varphi_\ell, \varphi_n \rangle \psi_n = \sum_n \frac{1}{s_n} \langle \varphi_\ell, T^*T\varphi_n \rangle \tag{6.7.2}$$

$$= \sum_n \langle T\varphi_\ell, \frac{1}{s_n} \varphi_n \rangle \tag{6.7.3}$$

$$= \sum_n \langle T\varphi_\ell, \psi_n \rangle \tag{6.7.4}$$

$$= T\varphi_\ell. \tag{6.7.5}$$

On the other hand, the series (6.7.1) is well defined (because it is normally convergent in $\mathcal{L}(\mathcal{H})$). It can also be verified that the formula is correct on $N(T)$.

Let us conclude with a few remarks.

1. We have a similar formula for T^* just by taking the adjoint of the formula for T:

$$T^* = \sum_n s_n \langle \cdot, \psi_n \rangle \varphi_n. \tag{6.7.6}$$

2. We have

$$\varphi_n = \frac{1}{s_n} T^* \psi_n.$$

3. ψ_n is an eigenfunction of TT^* associated with the eigenvalue s_n^2. We have, in fact,

$$(TT^*)\psi_n = \frac{1}{s_n} TT^*T\varphi_n = s_n(T\varphi_n) = s_n^2 \psi_n.$$

Moreover, s_n^2 has the same multiplicity for TT^* and T^*T.

4. s_n is an eigenvalue of $|T| = \sqrt{T^*T}$ and is called a singular value of T.

6.8 Trace class operators and Hilbert–Schmidt operators

6.8.1 The trace for nonnegative self-adjoint operators

Definition 6.24 (Trace of a nonnegative operator.) Let T be a bounded non-negative self-adjoint operator on a separable Hilbert space $\mathcal{H}$. The sum of the series (finite or not)[2] with general term $\langle T\varphi_n,\ \varphi_n \rangle$, where (φ_n) is a Hilbertian basis of $\mathcal{H}$, is independent of the basis. It is called the trace of T and denoted by Tr T,

$$\text{Tr } T = \sum_{1}^{+\infty} \langle T\varphi_n,\ \varphi_n \rangle.$$

This trace coincides with the usual trace in the case of matrices and keeps most of the properties established in that case. Nevertheless, in order to have a coherent definition, we need first to check that the definition does not depend on the choice of the orthonormal basis.

We have seen that if T is nonnegative, then T admits a unique nonnegative square root denoted by S, i.e., a nonnegative operator $S \in \mathcal{L}(\mathcal{H})$ such that $S^2 = T$. If in addition T is a compact operator, it is easy to define it in an orthonormal basis of eigenfunctions (e_j) of T, so that $Te_j = \lambda_j e_j$, by writing $Se_j = \lambda_j^{1/2} e_j$.

We observe, incidentally, that Tr T is equal to $\sum_j \lambda_j$ (where the eigenvalues are counted with multiplicity).

Proof of the independence

$$
\begin{aligned}
\sum_n \langle T\varphi_n,\ \varphi_n \rangle &= \sum_n \|S\varphi_n\|^2 \\
&= \sum_{n,m} |\langle S\varphi_n,\ \psi_m \rangle|^2 \\
&= \sum_{n,m} |\langle \varphi_n,\ S\psi_m \rangle|^2 = \sum_m \|S\psi_m\|^2 \\
&= \sum_m \langle T\psi_m,\ \psi_m \rangle
\end{aligned}
$$

To justify the interchange of the sums, we use Fubini's theorem for double series (observing that all the terms are nonnegative). $\qquad\square$

Main properties

- If T_1 and T_2 are nonnegative, then Tr $(T_1 + T_2) = $ Tr $T_1 + $ Tr T_2.
- For $\lambda > 0$ and T nonnegative, Tr $(\lambda T) = \lambda$ Tr (T).
- Tr $(U^{-1}TU) = $ Tr (T) for any nonnegative T and any unitary operator U.
- If $0 \le T_1 \le T_2$, then $0 \le $ Tr $(T_1) \le $ Tr (T_2).

---

[2] This is no problem, because all the terms are nonnegative.

6.8.2 Trace class operators

Definition 6.25 (Trace class operator.) A bounded operator is trace class if the trace of $|T| = (T^*T)^{1/2}$ is finite. We denote by $\mathcal{L}_1$ the space of trace class operators.

Polar decomposition is useful for analyzing the properties of trace class operators.

Proposition 6.26 *(Polar decomposition.) If T is a bounded operator, then we can always write*

$$T = U\,|T|, \tag{6.8.1}$$

where U is a partial isometry (i.e., an isometry on $N(T)^{\perp}$) with kernel $N(T)$. Moreover, we have

$$UT^* = TU^*,\ UT^*TU^* = TT^*,\ U\,|T|\,U^* = |T^*|,\ |T| = U^*T. \tag{6.8.2}$$

Proof For any x, we have

$$\|\,|T|\,x\|^2 = \|Tx\|^2. \tag{6.8.3}$$

Using (6.8.3), we define an operator U on $R(|T|)$ by

$$U|T|x = Tx,$$

with values in $R(T)$. U preserves the norm and can be extended to $\overline{R(|T|)}$ as an isometric map on $\overline{R(|T|)}$ into $\overline{R(T)}$. We can now extend U by 0 on $\overline{R(|T|)}^{\perp}$. Then it is clear that $N(T) = N(U)$. We have, in fact, $N(T) = N(|T|) = R(|T|)^{\perp}$ and $R(|T|)^{\perp} \subset N(U)$ by construction.

Conversely, if for $y = |T|x$, $Uy = 0$, we then obtain $Tx = 0$ and $y = 0$ by (6.8.3). We then immediately obtain (6.8.1).

Let us prove the four formulas of (6.8.2). The first one follows from

$$T^* = |T|\,U^*. \tag{6.8.4}$$

For the second formula, we observe that

$$UT^*TU^* = TU^*TU^* = TU^*U\,|T|\,U^*.$$

Then, we use the fact that U^*U is the identity on $R(|T|)$:

$$U^*U\,|T| = |T|, \tag{6.8.5}$$

which is a consequence of the fact that U is a partial isometry. Hence we obtain

$$TU^*U\,|T|\,U^* = T\,|T|\,U^* = (U\,|T|)(|T|\,U^*) = TT^*,$$

using (6.8.4).

For the third formula, we consider the square $U\,|T|\,U^*U\,|T|\,U^*$ and, using (6.8.5),

$$U\,|T|\,U^*U\,|T|\,U^* = U\,|T|^2\,U^* = TT^*.$$

For the last formula, we can again use (6.8.5).   □

Proposition 6.27 $\mathcal{L}_1$ *is a bilateral ideal in* $\mathcal{L}(\mathcal{H})$ *that is stable with respect to taking the adjoint.*

In the standard definition of a bilateral ideal in an algebra (here the product in the algebra is the composition of the operators), it is stated in particular that when an element of $\mathcal{L}_1$ is composed on the left or on the right with an element of $\mathcal{L}(\mathcal{H})$, one remains in $\mathcal{L}_1$.

Using polar decomposition, we can write

$$T_1 = U_1\,|T_1|, \quad T_2 = U_2\,|T_2|, \quad (T_1 + T_2) = U\,|T_1 + T_2|$$

and then compute, for a given $N \in \mathbb{N}^*$, the finite sum $\sum_1^N \langle \phi_n,\ |T_1 + T_2|\,\phi_n \rangle$:

$$
\begin{aligned}
\sum_{n=1}^{N} & \langle \phi_n,\ |T_1 + T_2|\,\phi_n \rangle \\
&= \sum_1^N \langle \phi_n,\ U^*(T_1 + T_2)\phi_n \rangle \\
&= \sum_1^N \langle \phi_n,\ U^*U_1\,|T_1|\,\phi_n \rangle + \sum_1^N \langle \phi_n,\ U^*U_2\,|T_2|\,\phi_n \rangle \\
&\leq \sum_1^N |\langle |T_1|^{1/2}\,U_1^*U\phi_n,\ |T_1|^{1/2}\,\phi_n \rangle| \\
&\quad + \sum_1^N |\langle |T_2|^{1/2}\,U_2^*U\phi_n,\ |T_2|^{1/2}\,\phi_n \rangle| \\
&\leq \left(\sum_1^N \|\,|T_1|^{1/2}\,U_1^*U\phi_n\|^2 \right)^{1/2} \left(\sum_1^N \|\,|T_1|^{1/2}\,\phi_n\|^2 \right)^{1/2} \\
&\quad + \left(\sum_1^N \|\,|T_2|^{1/2}\,U_2^*U\phi_n\|^2 \right)^{1/2} \left(\sum_1^N \|\,|T_2|^{1/2}\,\phi_n\|^2 \right)^{1/2} \\
&\leq \left(\sum_1^N \|\,|T_1|^{1/2}\,U_1^*U\phi_n\|^2 \right)^{1/2} (\mathrm{Tr}\,|T_1|)^{1/2} \\
&\quad + \left(\sum_1^N \|\,|T_2|^{1/2}\,U_2^*U\phi_n\|^2 \right)^{1/2} (\mathrm{Tr}\,|T_2|)^{1/2} \\
&\leq \left(\mathrm{Tr}\,(U^*U_1\,|T_1|\,U_1^*U) \right)^{1/2} (\mathrm{Tr}\,|T_1|)^{1/2}
\end{aligned}
$$

$$+ \left(\text{Tr} \, \left(U^* U_2 \, |T_2| \, U_2^* U \right) \right)^{1/2} \left(\text{Tr} \, |T_2| \right)^{1/2}$$
$$\leq \text{Tr} \, (|T_1|) + \text{Tr} \, (|T_2|) < +\infty. \tag{6.8.6}$$

On the way, we have used the fourth equality in (6.8.2), the Cauchy–Schwarz inequality, and the property that if U is a partial isometry and if S is nonnegative, we have

$$\text{Tr} \, U^* S U \leq \text{Tr} \, S. \tag{6.8.7}$$

In conclusion, we deduce from the above estimates that the series with general term $\langle \phi_n, |T_1 + T_2| \phi_n \rangle$ is convergent and its sum is, by definition, $\text{Tr} \, |T_1 + T_2|$.

The operator $(T_1 + T_2)$ is trace class (we have not yet defined the trace, but this will be done in Section 6.8.4), and we have also obtained

$$\text{Tr} \, |T_1 + T_2| \leq \text{Tr} \, |T_1| + \text{Tr} \, |T_2|. \tag{6.8.8}$$

Remark 6.28 The inequality (6.8.8) is not a consequence of an estimate of the type $|T_1 + T_2| \leq |T_1| + |T_2|$. This estimate is wrong! A simple example is the following:

$$T_1 = \begin{pmatrix} 1 & 1 \\ 1 & 1 \end{pmatrix}, \ T_2 = \begin{pmatrix} 0 & 0 \\ 0 & -2 \end{pmatrix}.$$

It can be verified that

$$|T_1 + T_2| = \begin{pmatrix} \sqrt{2} & 0 \\ 0 & \sqrt{2} \end{pmatrix} \ \text{and} \ |T_1| + |T_2| = \begin{pmatrix} 1 & 1 \\ 1 & 3 \end{pmatrix}$$

It is then easy to find a counterexample using the vector $(1, 0)$.

At this point, we have simply shown that $\mathcal{L}_1$ is a vector space. We would need to do more work to show that it is an ideal in $\mathcal{L}(H)$. For the property of stability when the adjoint is taken, we can observe that we have shown that $|T^*| = U \, |T| \, U^*$, and we can then use the fact that $\mathcal{L}_1$ is an ideal.

Proposition 6.29 *Any trace class operator is compact.*

Proof By (6.8.1), it is enough to show that $|T|$ is compact. But if $|T|$ has a finite trace and is not of finite rank, the decreasing sequence of its eigenvalues (s_n) is infinite and tends to 0. If Π_n is the projector associated with the eigenvalues of $|T|$ that are larger than s_n, it is then clear that $\Pi_n |T|$ is of finite rank and that $\Pi_n |T|$ converges to $|T|$ in $\mathcal{L}(\mathcal{H})$. Hence $|T|$ is compact. $\square$

Conversely, it can be shown that T is a compact operator if the series of its singular values (s_i) (i.e., the positive square roots of the nonzero eigenvalues of T^*T) is convergent; then T is trace class. We can in fact find a basis of eigenvectors of T^*T such that

$$|T| = \sum_i s_i \langle \cdot, \psi_i \rangle \psi_i,$$

and we have $T \in \mathcal{L}_1$ if and only if $\sum_i s_i < +\infty$.

Proposition 6.30 *The map* $T \mapsto \mathrm{Tr}(|T|)$ *is a norm on* $\mathcal{L}_1$*, which becomes a Banach space with this norm. Moreover, we have*

$$||T||_{\mathcal{L}(\mathcal{H})} \le ||T||_{\mathcal{L}_1} := \mathrm{Tr}\,|T|. \tag{6.8.9}$$

Proof To compare the two norms, we can first use polar decomposition to reduce the problem to the case of a nonnegative operator T. But in this case this is simply the trivial inequality

$$\sup_j \lambda_j \le \sum_j \lambda_j.$$

We also note that $\mathrm{Tr}\,|T| = 0$ implies that $T^*T = 0$, and hence $T = 0$. It remains to show that $\mathcal{L}_1$ is complete. We have to show that any Cauchy sequence (T_n) in $\mathcal{L}_1$, i.e., such that

$$\forall \epsilon > 0, \ \exists N \text{ s.t. } \forall n, p \ge N, \ ||T_n - T_p||_{\mathcal{L}_1} < \epsilon,$$

is convergent.

The first step is to observe that (T_n) is also a Cauchy sequence in $\mathcal{L}(\mathcal{H})$. Consequently, there exists T in $\mathcal{L}(\mathcal{H})$ such that $T_n \to T$ in $\mathcal{L}(\mathcal{H})$. It remains to show that T is trace class and that T_n tends to T in $\mathcal{L}_1$.

The second step is to observe that a result of the assumption that T_n is bounded in $\mathcal{L}_1$ is that there exists $M > 0$ such that

$$\mathrm{Tr}\,|T_n| \le M, \ \forall n \in \mathbb{N}.$$

As above, following the proof of the triangle inequality but with $T_1 = T - T_n$ and $T_2 = T_n$), we obtain, for any m, n, and any orthonormal basis $(\phi_k)_k$,

$$\sum_{k=1}^m \langle |T| \phi_k, \phi_k \rangle \ \le m \, ||T - T_n||_{\mathcal{L}(\mathcal{H})} + M.$$

For any m, we can find n such that

$$m \, |||T - T_n| \, ||_{\mathcal{L}(\mathcal{H})} \le 1.$$

Hence we obtain, for any m, the inequality

$$\sum_{k=1}^{m} \langle |T| \phi_k, \phi_k \rangle \le M + 1,$$

which shows that $T \in \mathcal{L}_1$.

Finally, we show that T_n tends to T in $\mathcal{L}_1$. We first observe that for all m, n, and p, we have (see (6.8.6) above)

$$\sum_{k=1}^{m} \langle |T - T_n| \phi_k, \phi_k \rangle \le m \, \|T - T_p\|_{\mathcal{L}(\mathcal{H})} + \sum_{k=1}^{+\infty} \langle |T_n - T_p| \phi_k, \phi_k \rangle.$$

We can then use the property that T_n is a Cauchy sequence in $\mathcal{L}_1$ and is convergent to T in $\mathcal{L}(\mathcal{H})$ to obtain the result (first take $p \to +\infty$ and then $m \to +\infty$). $\qquad\square$

Finally, it is not too difficult, following what has been done for compact operators, to show that the finite-rank operators are dense in $\mathcal{L}_1$.

6.8.3 Hilbert–Schmidt operators

Definition 6.31 An operator T is Hilbert–Schmidt if T^*T is trace class.

It can be seen immediately that the set of the Hilbert–Schmidt operators is a vector space. The trace class condition reads

$$\sum_{k} \|T\phi_k\|^2 < +\infty,$$

for an orthonormal basis (φ_k), and we have seen that this quantity is independent of the choice of the orthonormal basis.

Proposition 6.32 *The map* $(T_1, T_2) \mapsto \mathrm{Tr}\,(T_2^*T_1)$ *is well defined and gives a scalar product on the space of the Hilbert–Schmidt operators, which is denoted by* $\mathcal{L}_2$. *With this scalar product,* $\mathcal{L}_2$ *is a Hilbert space.*

A walk through the proof By the Cauchy–Schwarz inequality, we can deduce immediately the absolute convergence of the series with general term $\langle T_1\phi_n, T_2\phi_n \rangle$. We can then easily verify that $\mathcal{L}_1 \subset \mathcal{L}_2$ and that $\mathcal{L}_2 \subset \mathcal{K}(\mathcal{H})$ (the space of the compact operators on $\mathcal{H}$). We also have

$$T \in \mathcal{L}_2 \text{ if and only if } \sum_{n} \lambda_n^2 < +\infty, \qquad (6.8.10)$$

where the λ_j are the singular values of T.

Moreover, it can be shown that $\mathcal{L}_2$ is a bilateral ideal in $\mathcal{L}(\mathcal{H})$. Finally, we have

$$\|T_1 T_2\|_{\mathcal{L}_1} \leq \|T_1\|_{\mathcal{L}_2} \| T_2\|_{\mathcal{L}_2}. \tag{6.8.11}$$

The following inequalities are also useful:

$$\|T_1 T_2\|_{\mathcal{L}_1} \leq \|T_1\|_{\mathcal{L}(\mathcal{H})} \|T_2\|_{\mathcal{L}_1}, \tag{6.8.12}$$

$$\|T_1 T_2\|_{\mathcal{L}_2} \leq \|T_1\|_{\mathcal{L}(\mathcal{H})} \|T_2\|_{\mathcal{L}_2}, \tag{6.8.13}$$

$$\|T_1 T_2\|_{\mathcal{L}_1} \leq \|T_1\|_{\mathcal{L}_1} \|T_2\|_{\mathcal{L}(\mathcal{H})}, \tag{6.8.14}$$

and

$$\|T_1 T_2\|_{\mathcal{L}_2} \leq \|T_1\|_{\mathcal{L}_2} \|T_2\|_{\mathcal{L}(\mathcal{H})}, \tag{6.8.15}$$

These inequalities show that the natural operations of composition are continuous.

6.8.4 Trace of a trace class operator

The title of this section may be surprising, but we have not yet defined the trace of an operator in $\mathcal{L}_1$.

Proposition 6.33 *If $T \in \mathcal{L}_1$, the series $\langle T\varphi_n, \varphi_n \rangle$ is absolutely convergent and its sum is independent of the chosen basis.*

Proof We can always write

$$T = (U |T|^{1/2})|T|^{1/2}.$$

Then the sum of the series appears as the scalar product of $|T|^{1/2}$ and $(U |T|^{1/2})^*$ in $\mathcal{L}_2$. $\square$

This leads to the following definition.

Definition 6.34 (Trace in $\mathcal{L}_1$.) For $T \in \mathcal{L}_1$, we set

$$\operatorname{Tr} T = \sum_n \langle T\varphi_n, \varphi_n \rangle.$$

We have the following proposition.

Proposition 6.35 *The trace map has the following properties:*

1. It is a continuous linear form on $\mathcal{L}_1$ and satisfies

$$\operatorname{Tr} T^* = \overline{\operatorname{Tr} T}.$$

2. For all $T_1 \in \mathcal{L}_1$ and $T_2 \in \mathcal{L}(H)$, we have

$$\mathrm{Tr}(T_1 T_2) = \mathrm{Tr}(T_2 T_1)$$

and

$$|\mathrm{Tr}(T_1 T_2)| \leq ||T_2||_{\mathcal{L}(\mathcal{H})} ||T_1||_{\mathcal{L}_1}.$$

6.8.5 Hilbert–Schmidt operators on $L^2(X, d\mu)$

If μ is a measure on a locally compact topological space, there is a strong link (actually an isometry) between the Hilbert–Schmidt operators on $L^2(X, d\mu)$ (where $L^2(X, d\mu)$ is the space of μ-measurable functions that are square integrable with respect to the same measure) and the operators whose distribution kernel is in $L^2(X \times X, d\mu \otimes d\mu)$.

To simplify the treatment, we consider the case of $L^2(\mathbb{R}^n)$ with the standard Lebesgue measure. If $K(x, y) \in \mathbb{S}(\mathbb{R}^n \times \mathbb{R}^n)$, then the operator $\mathbb{K}$ with the distribution kernel K defined by

$$C_0^\infty(\mathbb{R}^n) \ni f \mapsto (\mathbb{K}f) := \int K(x, y) f(y) \, dy$$

admits a unique continuous extension (also denoted by $\mathbb{K}$) from L^2 into L^2, and we have

$$||\mathbb{K}||_{\mathcal{L}(L^2)} \leq ||K||_{L^2}.$$

Proposition 6.36 *The operator $\mathbb{K}$ belongs to $\mathcal{L}_2$, and the map that associates the operator $\mathbb{K}$ with K is an isometry from $L^2(\mathbb{R}^n \times \mathbb{R}^n)$ onto $\mathcal{L}_2$.*

Proof Let $(\phi_q)_q$ be an orthonormal basis in $L^2(\mathbb{R}^n)$. Let us first assume that K is in $\mathcal{S}(\mathbb{R}^n \times \mathbb{R}^n)$. Observing that the double sequence $\phi_{q,m} : (x, y) \mapsto \phi_q(x)\overline{\phi_m(y)}$ $((q, m) \in (\mathbb{N}^*)^2)$ is an orthonormal basis of $L^2(\mathbb{R}^n \times \mathbb{R}^n)$, we can write

$$K(x, y) = \sum_{m,q} K_{q,m} \phi_q(x)\overline{\phi_m(y)}.$$

We can then write (using finite sums to justify our argument)

$$\mathbb{K}\phi_m = \sum_q K_{q,m}\phi_q,$$

which implies

$$||\mathbb{K}\phi_m||^2 = \sum_q |K_{q,m}|^2.$$

This leads to

$$\sum_m ||\mathbb{K}\phi_m||^2 = \sum_{q,m} |K_{q,m}|^2 = ||K||^2_{L^2}.$$

Hence $\mathbb{K}$ is a Hilbert–Schmidt operator and the other assertion is proven. $\quad\square$

Remark 6.37 In the case of a Hermitian integral kernel, i.e., a kernel satisfying $K(x, y) = \overline{K(y, x)}$, the associated operator is self-adjoint. Considering an orthonormal basis $(\phi_n)_{n \in \mathbb{N}^*}$ of eigenfunctions of K, we obtain

$$\mathbb{K} = \sum_j \lambda_j \langle \cdot, \phi_j \rangle \phi_j,$$

and we have

$$\sum_j \lambda_j^2 = ||K||^2_{L^2} = ||\mathbb{K}||^2_{\mathcal{L}_2}.$$

Remark 6.38 Note that this isometry is actually surjective on $\mathcal{L}_2$. We can verify that the finite-rank operators are in the range of the isometry, and then conclude by observing that these finite-rank operators are dense in $\mathcal{L}_2$.

6.9 Krein–Rutman theorem

This theorem is the natural extension of the Perron–Frobenius theorem for matrices to an infinite-dimensional situation. First, we state a definition.

Definition 6.39 Let A be a bounded nonnegative operator on a Hilbert space $\mathcal{H} = L^2(X, d\nu)$, where (X, ν) is a measured space. Then we say that A has a positive kernel if, for each choice of a nonnegative function $\theta \in \mathcal{H}$ ($||\theta|| \neq 0$), we have

$$0 < A\theta$$

almost everywhere.

It can be seen immediately that the transfer operator satisfies this condition.

The theorem generalizing the Perron–Frobenius theorem is then the following.

Theorem 6.40 *(Krein–Rutman theorem.) Let A be a bounded nonnegative compact symmetric operator on $\mathcal{H}$ with a positive kernel, and let[3] $||A|| = \lambda$ be the largest eigenvalue of A. Then λ has multiplicity 1, and the corresponding eigenfunction u_λ can be chosen to be a positive function.*

---

[3] See Proposition 6.20.

Proof Since A maps real functions into real functions, we may assume that u_λ is real. We now prove that

$$\langle Au_\lambda, u_\lambda \rangle \leq \langle A|u_\lambda|, |u_\lambda| \rangle.$$

This is an immediate consequence of the positivity of the kernel. We just write

$$u_\lambda = u_\lambda^+ - u_\lambda^-$$

and

$$|u_\lambda| = u_\lambda^+ + u_\lambda^-,$$

and the above inequality is then a consequence of

$$\langle Au_\lambda^+, u_\lambda^- \rangle \geq 0$$

and

$$\langle u_\lambda^+, Au_\lambda^- \rangle \geq 0.$$

We then obtain

$$\lambda||u_\lambda||^2 = \langle Au_\lambda, u_\lambda \rangle \leq \langle A|u_\lambda|, |u_\lambda| \rangle \leq ||A|| \, ||u_\lambda||^2 = \lambda||u_\lambda||^2.$$

This implies

$$\langle Au_\lambda, u_\lambda \rangle = \langle A|u_\lambda|, |u_\lambda| \rangle.$$

This equality means

$$\langle u_\lambda^+, Au_\lambda^- \rangle + \langle u_\lambda^-, Au_\lambda^+ \rangle = 0.$$

We then obtain a contradiction unless $u_\lambda^+ = 0$ or $u_\lambda^- = 0$. We can assume $u_\lambda \geq 0$, and the assumption gives again

$$0 < \langle \theta, Au_\lambda \rangle = \lambda\langle \theta, u_\lambda \rangle,$$

for any nonnegative θ. This gives

$$u_\lambda \geq 0 \quad \text{a.e.}$$

But

$$u_\lambda = \lambda^{-1} Au_\lambda,$$

and this gives

$$u_\lambda > 0 \quad \text{a.e.}$$

Finally, if there were two linearly independent eigenfunctions v_λ and u_λ corresponding to λ, we would obtain the same property for v_λ by considering as a new Hilbert space the orthogonal of u_λ in $\mathcal{H}$. But it is impossible to have two orthogonal vectors that are positive. □

6.10 Notes

The material of this section is basic and is described in many monographs devoted to spectral theory. For trace class operators and Hilbert–Schmidt operators, readers are referred to, for example, the books by Lévy-Bruhl [Le-Br, section 7], Robert [Ro], and Simon [Si2], where complete proofs are given.

In the analysis of the index of Fredholm operators, we have used the technique of the so-called Grushin's problem. Readers are referred to the exposition by Sjöstrand and Zworski [SjZw] for a presentation of this method and its applications in spectral theory.

It is interesting to mention an important theorem due to Lidskii that states that the trace of a trace class operator is the sum of its eigenvalues. A proof can be found in [Si2], for example.

6.11 Exercises

Exercise 6.1 Let $\alpha \in [0, 1]$. Let p and q be integers that are mutually prime. Analyze the spectrum Σ_α of the operator H_α defined on $\ell^2(\mathbb{Z})$ by

$$(H_\alpha u)_n = \frac{1}{2}(u_{n-1} + u_{n+1}) + \cos 2\pi \left(\frac{p}{q}n + \alpha\right) u_n, \ n \in \mathbb{Z}.$$

In order to make the analysis easier, you may assume (this is a particular case of the so-called Floquet theory) that $\Sigma = \cap_{\theta \in [0,1]} \Sigma_\theta$, where Σ_θ is the spectrum of H_α reduced to the space of the u's in ℓ^∞ such that $u_{n+q} = \exp 2i\pi\theta \, u_n$, for $n \in \mathbb{Z}$. This operator plays an important role in solid state physics and is called the Harper operator (whose spectrum as a function of $\beta = p/q$ describes the celebrated Hofstadter butterfly).

We now replace the rational number p/q by an irrational number β. So, consider the operator $H_{\beta,\alpha} := \frac{1}{2}(\tau_{+1} + \tau_{-1}) + \cos 2\pi(\beta \cdot + \alpha)$ on $\ell^2(\mathbb{Z}, \mathbb{C})$, where τ_k $(k \in \mathbb{Z})$ is the operator defined on $\ell^2(\mathbb{Z}, \mathbb{C})$ by $(\tau_k u)_n = u_{n-k}$. Show that if $\beta \notin \mathbb{Q}$, then the spectrum of $H_{\beta,\alpha}$ is independent of α. For this, we suggest that you first prove that $H_{\beta,\alpha}$ is unitarily equivalent to $H_{\beta,\alpha+k\beta}$ for any

$k \in \mathbb{Z}$. Secondly, use the density of the set $\{\alpha + \beta\mathbb{Z} + \mathbb{Z}\}$ in $\mathbb{R}$. Finally, use the inequality

$$\|H_{\beta,\alpha} - H_{\beta,\alpha'}\| \leq 2\pi |\alpha - \alpha'|.$$

Exercise 6.2 (Inspired by P. Gérard and S. Grellier.) Let $c = (c_n)_{n\in\mathbb{N}}$ be a real sequence, and consider the associated operator Γ_c on $\ell^2(\mathbb{N})$ defined by

$$(\Gamma_c x)_n = \sum_{p\geq 0} c_{n+p} x_p.$$

Show that Γ_c is Hilbert–Schmidt if and only if $\sum_{n\in\mathbb{N}} n |c_n|^2 < +\infty$.

Exercise 6.3 (See Example 6.12.)

(a) Let g be a continuous function on $\mathbb{R}$ such that $g(0) = 0$. Analyze the convergence of the sequence $(g(t)u_n(t))_{n\geq 1}$ in $L^2(\mathbb{R})$, where $u_n(t) = \sqrt{n}\chi(nt)$ and χ is a C^∞ function with compact support.

(b) Let $f \in C^0([0, 1]; \mathbb{R})$. Let T_f be the operator of multiplication by f defined on $L^2(]0, 1[)$ by $u \mapsto T_f u = fu$. Determine the spectrum of T_f. Discuss as a function of f the possible existence of eigenvalues. Determine the essential spectrum of T_f.

Exercise 6.4 Let T be the operator defined for $f \in \mathcal{H} = L^2(]0, 1[; \mathbb{R})$ by

$$(Tf)(x) = \int_0^x f(t)\,dt$$

(see Exercise 5.3). Does T have eigenvalues? Determine the spectrum of T.

Exercise 6.5 (After Effros, and Avron, Seiler, and Simon.) Let P and Q be two self-adjoint projectors in a Hilbert space $\mathcal{H}$.

(a) Assume that $A = P - Q$ is compact. Show that if $\lambda \neq \pm 1$ is in the spectrum, then $-\lambda$ is in the spectrum with the same multiplicity. For this, we suggest that you show first that with $B = I - P - Q$,

$$A^2 + B^2 = I, \quad AB + BA = 0.$$

(b) Assume now, in addition, that A is trace class. Compute the trace and show that it is an integer.

Exercise 6.6 Let Ω be a nonempty open subset in $\mathbb{R}^d$, and consider the multiplication operator on $L^2(\mathbb{R}^d)$ defined by multiplication by χ_Ω, where χ_Ω is equal to 1 in Ω and 0 outside. Determine the spectrum, the essential spectrum, and the discrete spectrum.

Exercise 6.7 Let $V \in \mathcal{S}(\mathbb{R}^d)$ be positive, and consider the operator

$$T := (-\Delta + 1)^{-1/2} V (-\Delta + 1)^{-1/2}.$$

(a) Explain how to define $(-\Delta + 1)^{-1/2}$ as an operator on $L^2(\mathbb{R}^d)$.
(b) Show that T is a bounded, self-adjoint, positive, compact operator on $L^2(\mathbb{R}^d)$.
(c) Discuss the injectivity of T as a function of V.
(d) Establish a link with the search for pairs (u, μ) in $H^2(\mathbb{R}^d) \times \mathbb{R}^+$ such that

$$(-\Delta + 1 - \mu V)u = 0.$$

Exercise 6.8 Consider on $\ell^2(\mathbb{N})$ the right-shift operator defined by $(\tau_r u)_j = u_{j-1}$ for $j \geq 1$ and by $(\tau_r u)_0 = 0$. Show that the spectrum is the whole unit disk. Determine τ_r^*.

Hint: First show that, for any $z \in \mathbb{C}$, $\tau_r - z$ is injective. If $z = 0$, show that τ_r is not surjective. For $z \neq 0$, show that one can solve $z \in \mathbb{C}$, $\tau_r u - zu = f$, with $f_j = 1/z^{j+1}$ if and only if $|z| > 1$.

Exercise 6.9 For $x \in [0, 1]$, consider

$$A(x) = \begin{pmatrix} x + 1 & -x \\ -x & x + 1 \end{pmatrix}.$$

Determine the spectrum of the associated multiplication operator on $L^2(]0, 1[; \mathbb{R}^2)$. Does it have eigenvalues? What is the essential spectrum?

Exercise 6.10 Let $\mathbf{K}_f$ be the operator of convolution by f on $L^2(\mathbb{S}^1)$. Assume that f is real and in $L^1(\mathbb{S}^1)$.

(a) Show that $\theta \mapsto 1$ is an eigenfunction, and give the associated eigenvalue.
(b) Assume that $f \geq 0$ (a.e.). Compute the norm of $\mathbf{K}_f$.
(c) Show that if f is in $L^1(\mathbb{S}^1)$, then $\mathbf{K}_f$ is a compact operator.

Exercise 6.11 Let $\mathcal{F}$ be the Fourier transform on $L^2(\mathbb{R})$ (see (2.1.5)).

(a) Show that the spectrum of $\mathcal{F}$ is contained in the unit circle of $\mathbb{C}$.
(b) Show that $\mathcal{F}^4 = I$. Deduce a simple inclusion for the set of the eigenvalues of $\mathcal{F}$.
(c) Show that if $L = d/dx - x$, then

$$L\mathcal{F} = i\mathcal{F}L,$$

in $\mathcal{L}(\mathcal{S}(\mathbb{R}))$. Similarly, verify that with $L^* = -d/dx - x$,

$$L^*\mathcal{F} = -i\mathcal{F}L^*.$$

(d) Verify that $\phi_1(x) = \pi^{-1/2}\exp-(x^2/2)$ is an eigenfunction of $\mathcal{F}$.
(e) Show that the $\phi_n(x) := L^{n-1}\phi_1$ $(n \geq 1)$ are also eigenfunctions, and determine the corresponding eigenvalues.
(f) Determine $\sigma(\mathcal{F})$.
(g) Show that

$$\mathcal{F} = e^{(i\pi/4)(D_x^2+x^2-1)}.$$

7

Applications to statistical mechanics and partial differential equations

In this chapter, we return to some of our previous examples and present a deeper analysis. The first example comes from statistical mechanics (continuous models). The other examples describe cases where the operators considered are unbounded but where one can return, by considering the inverse, to the spectral theory of compact self-adjoint operators.

7.1 A problem in statistical mechanics

7.1.1 The transfer operator

The transfer operator is the operator associated with the kernel

$$K_t(x, y) = \exp - \frac{V(x)}{2} \exp -t \, |x - y|^2 \exp - \frac{V(y)}{2}, \qquad (7.1.1)$$

where $t > 0$, and V is a $C^\infty(\mathbb{R})$ function such that

$$\int_\mathbb{R} \exp -V(x) \, dx < +\infty.$$

The L^2-boundedness of operators with an integral kernel is very often proven using the following lemma.

Lemma 7.1 *(Schur's lemma.) Let $\mathbf{K}$ be an operator associated with an integral kernel K, that is, a function $(x, y) \mapsto K(x, y)$ on $\mathbb{R}^m \times \mathbb{R}^m$, satisfying*

$$M_1 := \sup_{x \in \mathbb{R}^m} \int_{\mathbb{R}^m} |K(x, y)| \, dy < +\infty,$$

$$M_2 := \sup_{y \in \mathbb{R}^m} \int_{\mathbb{R}^m} |K(x, y)| \, dx < +\infty. \qquad (7.1.2)$$

Then $\mathbf{K}$, initially defined for $u \in C_0^\infty(\mathbb{R}^m)$ by

$$(\mathbf{K}u)(x) = \int_{\mathbb{R}^m} K(x, y) u(y) \, dy,$$

can be extended as a continuous linear operator in $\mathcal{L}(L^2(\mathbb{R}^m))$ (still denoted by $\mathbf{K}$ or T_K), whose norm satisfies

$$\|\mathbf{K}\| \leq \sqrt{M_1 M_2}. \tag{7.1.3}$$

Proof By the Cauchy–Schwarz inequality, we have

$$|\mathbf{K}u(x)|^2 \leq \int |K(x, y)| \, |u(y)|^2 \, dy \int |K(x, y)| \, dy$$
$$\leq M_1 \int |K(x, y)| \, |u(y)|^2 \, dy.$$

By integrating with respect to x and using Fubini's theorem, we then obtain the result. □

In our case, the operator T_K is actually a Hilbert–Schmidt operator (see Section 6.8.5). We recall that it can be proved, using the Cauchy–Schwarz inequality, that

$$|\mathbf{K}u(x)|^2 \leq \int |u(y)|^2 \, dy \int |K(x, y)|^2 \, dy,$$

and we obtain

$$\|T_K\| \leq \|K\|_{L^2(\mathbb{R}^m \times \mathbb{R}^m)}. \tag{7.1.4}$$

Hence T_K is a compact operator. Consequently, it has a sequence of eigenvalues λ_n tending to 0 as $n \to +\infty$. Let us show the following lemma.

Lemma 7.2 *The transfer operator is positive and injective.*

Proof Let $u \in L^2(\mathbb{R})$. We can write,[1] with

$$\phi(x) = \exp -\frac{V(x)}{2} u(x),$$

---

[1] If we make only the weak assumption that $e^{-V} \in L^1(\mathbb{R})$, it is better to start the proof of the positivity by considering u's in $C_0^\infty(\mathbb{R})$ and then to treat the general case by using the density of C_0^∞ in L^2.

$$\langle T_K u, u \rangle_{\mathcal{H}} = \int_{\mathbb{R}^2} \exp -t \, |x - y|^2 \phi(x) \overline{\phi(y)} \, dx \, dy.$$

Using the properties of the Fourier transform (see (2.1.7)) and of the convolution, we deduce

$$\langle T_K u, u \rangle_{\mathcal{H}} = c_t \int_{\mathbb{R}} \exp -\frac{|\xi|^2}{4t} |\widehat{\phi}(\xi)|^2 \, d\xi,$$

where $c_t > 0$ is a normalization constant.

Consequently, the spectrum is the union of a sequence of positive eigenvalues and its limit 0. T_K can be diagonalized in an orthonormal basis of eigenfunctions associated with positive eigenvalues. We emphasize that 0 is in the spectrum but is not an eigenvalue. $\qquad\square$

Theorem 6.18 states also that $\|T_K\|$ is the largest eigenvalue and is isolated. It is then a natural question to discuss the multiplicity of each eigenvalue, i.e., the dimension of each associated eigenspace. We shall see later that the largest eigenvalue is of multiplicity 1.

7.1.2 The physical origin of the problem

Our initial aim was to find a rather general approach to the estimation of the decay of correlations attached to "Gaussian-like" measures of the type

$$\exp -\Phi(X) \, dX \qquad (7.1.5)$$

on $\mathbb{R}^m$ with Φ. One proof of such an estimate (when Φ has a particular structure) is based on an analysis of the transfer matrix method, originally due to the physicists Kramers and Wannier. Here, we briefly present the technique for our toy model. We shall consider only the case where $d = 1$ and treat the periodic case, that is, the case where $\{1, \cdots, n\}$ is a representation of $\Lambda^{\text{per}} = \mathbb{Z}/n\mathbb{Z}$.

We consider the particular potential Φ defined by

$$\Phi^{(n)}(X) \equiv \Phi(X) \equiv \frac{1}{h} \left(\sum_{j=1}^{n} V(x_j) + \frac{|x_j - x_{j+1}|^2}{4} \right), \qquad (7.1.6)$$

where we adopt the convention that $x_{n+1} = x_1$ and h is possibly a semiclassical parameter, which is sometimes chosen equal to one if we are not interested in the "semiclassical" aspects. More generally, we could consider examples of the form

$$\Phi_h(X) = \frac{1}{h} \left(\sum_{j=1}^{n} \left(V(x_j) + I(x_j, x_{j+1}) \right) \right), \qquad (7.1.7)$$

where I is a symmetric "interaction" potential on $\mathbb{R} \times \mathbb{R}$. Let us mention, however, that the example (7.1.6) appears naturally in quantum field theory when the "lattice approximation" is introduced. For this special class of potentials, we shall show what can be learned from spectral theory. We shall present the "dictionary" between the properties of the measure $h^{-n/2} \cdot \exp -\Phi_h(X) \, dX$ on $\mathbb{R}^m$ and the spectral properties of the transfer operator $\mathbf{K_V}$ (which is also called the Kac operator for some particular models introduced by M. Kac), whose integral kernel is real and given on $\mathbb{R} \times \mathbb{R}$ by

$$K_V(x, y) = h^{-1/2} \exp -\frac{V(x)}{2h} \cdot \exp -\frac{|x - y|^2}{4h} \cdot \exp -\frac{V(y)}{2h}. \qquad (7.1.8)$$

By an integral kernel (or distribution kernel), we mean[2] a distribution in $\mathcal{D}'(\mathbb{R}^2)$ such that the operator is defined from $C_0^\infty(\mathbb{R})$ into $\mathcal{D}'(\mathbb{R})$ by the formula

$$\int_\mathbb{R} (\mathbf{K_V} u)(x) \, v(x) \, dx = \int_{\mathbb{R} \times \mathbb{R}} K_V(x, y) \, u(y) \, v(x) \, dx \, dy, \quad \forall u, \; v \in C_0^\infty(\mathbb{R}). \qquad (7.1.9)$$

This dictionary permits us to obtain interesting connections between estimates of the quotient μ_2/μ_1 of the largest and second largest eigenvalues of the transfer operator and some corresponding estimates that can be used to control the speed of convergence of thermodynamic quantities. In particular, this speed of convergence is exponentially rapid as $n \to +\infty$.

We know that when the operator $\mathbf{K}$ is compact, symmetric, and injective, there exists a decreasing (in modulus) sequence μ_j of eigenvalues tending to 0 and a corresponding sequence of eigenfunctions u_j, which can be normalized in order to obtain an orthonormal basis of $L^2(\mathbb{R})$. Moreover, the operator becomes the limit in norm of the family of operators $\mathbf{K}^{(N)}$ whose corresponding kernels are defined by

$$K^{(N)}(x, y) = \sum_{j=1}^N \mu_j \, u_j(x) \, u_j(y). \qquad (7.1.10)$$

We can compute the trace of a trace class operator directly in the following way.

We first observe that

$$\mathrm{Tr} \, \mathbf{K} = \lim_{N \to +\infty} \mathrm{Tr} \, \mathbf{K}^{(N)}. \qquad (7.1.11)$$

---

[2] Here we shall always consider much more regular kernels that are, in particular, continuous. So, the notation "$\int$" can be interpreted in the usual sense. In general, this means that the distribution kernel K_V is applied to the test function $(x, y) \mapsto u(x) \, v(y)$.

Then we observe that

$$\text{Tr } \mathbf{K}^{(N)} = \sum_{j=1}^{N} \mu_j = \int_{\mathbb{R}} K^{(N)}(x, x)\, dx. \qquad (7.1.12)$$

Consequently, we obtain the result that

$$\text{Tr } \mathbf{K} = \lim_{N \to +\infty} \int_{\mathbb{R}} K^{(N)}(x, x)\, dx. \qquad (7.1.13)$$

If we observe that $x \mapsto \sum_j |\mu_j|\, |u_j(x)|^2$ is in $L^1(\mathbb{R})$, then it is natural to hope (but this is not trivial!) that $x \mapsto K(x, x)$ is in L^1 and that

$$\text{Tr } \mathbf{K} = \int_{\mathbb{R}} K(x, x)\, dx. \qquad (7.1.14)$$

Let us sketch a proof of (7.1.14) in our particular case. Since $\mathbf{K}$ is positive and has an explicit kernel,

$$\exp -\frac{V(x)}{2} \, \exp -\mathcal{J}|x - y|^2 \, \exp -\frac{V(y)}{2},$$

one can find a Hilbert–Schmidt operator $\mathbf{L}$ satisfying $\mathbf{L}^\star \mathbf{L} = \mathbf{K}$. The kernel of $\mathbf{L}$ is given, for a suitable θ, by

$$L(x, y) = c_\theta \, \exp -\theta |x - y|^2 \exp -\frac{V(y)}{2}.$$

We note that in fact $c_\theta \exp -\theta\, |x - y|^2$ is the distribution kernel of $\exp t_\theta \Delta$ for a suitable $t_\theta > 0$.

We then observe that $\mathbf{L}$ is Hilbert–Schmidt and that $||\mathbf{L}||_{\text{H.S.}}^2 = \sum_j \mu_j = \text{Tr } \mathbf{K}$. Using the previously mentioned formula for the Hilbert–Schmidt norm and the property that

$$K(x, x) = \int L^\star(x, z)\, L(z, x)\, dz = \int L^\star(z, x)\, L(z, x)\, dz,$$

we obtain (7.1.14). Note that it is only when $\mathbf{K}$ is nonnegative that the finiteness of the right-hand side of (7.1.14) implies the trace class property.

In the case where the operator is Hilbert–Schmidt, that is, with a kernel K in L^2, then the operator is also compact and we have the identity

$$\sum_j \mu_j^2 = \int_{\mathbb{R} \times \mathbb{R}} |K(x, y)|^2\, dx\, dy. \qquad (7.1.15)$$

Let us look first at the thermodynamic limit. This means that we are interested in the limit $\lim_{n \to +\infty} (1/n) \ln \left(\int_{\mathbb{R}^n} \exp -\Phi(X) \, dX \right)$. We start from the decomposition

$$\exp -\Phi(X) = K_V(x_1, x_2) \cdot K_V(x_2, x_3) \cdots K_V(x_{n-1}, x_n) \cdot K_V(x_n, x_1),$$
$$(7.1.16)$$

and observe that

$$\int_{\mathbb{R}^n} \exp -\Phi(X) \, dX = \int_{\mathbb{R}} K_{V,n}(y, y) \, dy, \qquad (7.1.17)$$

where $K_{V,n}(x, y)$ is the distribution kernel of $(\mathbf{K_V})^n$. Our assumption about V permits us to see that $(\mathbf{K_V})^n$ is trace class, and we rewrite (7.1.17) in the form

$$\int_{\mathbb{R}^n} \exp -\Phi(X) \, dX = \mathrm{Tr} \left[(\mathbf{K_V})^n \right] = \sum_j \mu_j^n, \qquad (7.1.18)$$

where the μ_j are defined in (7.1.23).

We note also for future use that

$$K_{V,n}(x, y) = \sum_j \mu_j^n u_j(x) u_j(y). \qquad (7.1.19)$$

In particular, in the thermodynamic limit we obtain

$$\lim_{n \to \infty} \frac{\ln \int_{\mathbb{R}^n} \exp -\Phi(X) \, dX}{n} = \ln \mu_1. \qquad (7.1.20)$$

Moreover, the speed of the convergence is easily estimated by use of the following equation:

$$-\ln \left| \frac{\ln \int_{\mathbb{R}^n} \exp -\Phi(X) \, dX}{n} - \ln \mu_1 \right|$$
$$= -n \ln \left(\frac{\mu_2}{\mu_1} \right) - \ln k_2 + \ln n + \mathcal{O}(\exp -\delta_2 n), \quad (7.1.21)$$

where k_2 is the multiplicity of μ_2.

7.1.3 Application of the Krein–Rutman theorem

We observe now that the kernel $(x, y) \mapsto K_V(x, y)$ satisfies the condition

$$K_V(x, y) > 0, \ \forall x, y \in \mathbb{R}. \qquad (7.1.22)$$

In particular, it satisfies the assumptions of the Krein–Rutman theorem (Theorem 6.40) and, consequently, our positive operator $\mathbf{K_V}$ has a largest eigenvalue μ_1 equal to $\|\mathbf{K_V}\|$, which is simple and corresponds to a unique positive normalized eigenfunction, which we denote by u_1. Let μ_j be the sequence of eigenvalues, which we order as a decreasing sequence tending to 0:

$$0 \leq \mu_{j+1} \leq \mu_j \leq \cdots \leq \mu_2 < \mu_1. \tag{7.1.23}$$

We denote by u_j a corresponding orthonormal basis of eigenfunctions with

$$\mathbf{K_V} u_j = \mu_j \, u_j, \; \|u_j\| = 1. \tag{7.1.24}$$

Remark 7.3 In the case $t = 0$, we keep the positivity but lose the injectivity! The spectrum is easy to determine. Assuming for normalization that $\int \exp -V \, dx = 1$, we are in fact dealing with the orthonormal projector associated with the function $x \mapsto \exp -V(x)/2$. The real number 1 is a simple eigenvalue, and 0 is an eigenvalue whose corresponding eigenspace is infinite dimensional.

7.2 The Dirichlet realization, a model of an operator with a compact resolvent

We return to the Dirichlet realization $-\Delta_D$ of the Laplacian in a bounded regular open set Ω. We have seen in Sections 4.4.1 and 5.3.2 that $-\Delta_D$ and $-\Delta_D + Id$ have a compact inverse.

We can apply Theorem 6.21 to the operator $(-\Delta_D + Id)^{-1}$. We have seen that this operator is compact, and it is clearly injective (by construction). It has also been seen to be self-adjoint and nonnegative. Moreover, the norm of this operator is less than or equal to 1.

Consequently, there exists a sequence of distinct eigenvalues μ_n tending to 0 (with $0 < \mu_n \leq 1$) and a corresponding orthonormal basis of eigenfunctions such that $(\Delta_D + I)^{-1}$ is diagonalized. If $\phi_{n,j}$ ($j = 1, \ldots, k_n$) is a corresponding basis of eigenfunctions associated with μ_n, that is, if

$$(-\Delta_D + I)^{-1} \phi_{n,j} = \mu_n \phi_{n,j},$$

we first observe that $\phi_{n,j} \in D(-\Delta_D + I)$; hence $\phi_{n,j} \in H_0^1(\Omega) \cap H^2(\Omega)$ (if Ω is relatively compact with a regular boundary) and

$$-\Delta_D \, \phi_{n,j} = \left(\frac{1}{\mu_n} - 1 \right) \phi_{n,j}.$$

The function $\phi_{n,j}$ is consequently an eigenfunction of $-\Delta_D$ associated with the eigenvalue $\lambda_n = (1/\mu_n - 1)$.

Let us show, as one may easily guess, that this basis $\phi_{n,j}$ effectively permits the diagonalization of $-\Delta_D$.

We consider $u = \sum_{n,j} u_{n,j}\, \phi_{n,j}$ in the domain of $-\Delta_D$, and the scalar product $\langle -\Delta_D u,\, \phi_{m,\ell} \rangle_{\mathcal{H}}$. Using the self-adjoint character of $-\Delta_D$, we obtain

$$\langle -\Delta_D u,\, \phi_{m,\ell} \rangle_{\mathcal{H}} = \langle u,\, -\Delta_D\, \phi_{m,\ell} \rangle_{\mathcal{H}} = \lambda_m u_{m,\ell}.$$

Observing that $D(\Delta_D) = R(S^{-1})$, we obtain the result that the domain of $-\Delta_D$ is characterized by

$$D(-\Delta_D) = \left\{ u \in L^2 \mid \sum_{n,j} \lambda_n^2\, |u_{n,j}|^2 < +\infty \right\}.$$

Here we have used the property that for any N, we have the following identity:

$$\sum_{n \leq N} u_{n,j}\phi_{n,j} = S\left(\sum_{n \leq N} \lambda_n u_{n,j}\phi_{n,j} \right).$$

Consequently, we have given a meaning to the following diagonalization formula:

$$-\Delta_D = \sum_n \lambda_n \Pi_{E_n}, \tag{7.2.1}$$

where Π_{E_n} is the orthogonal projector on the eigenspace E_n associated with the eigenvalue λ_n.

We remark that the property that λ_n tend to $+\infty$ results from the property that the sequence μ_n tends to 0.

Let us also prove the following lemma.

Lemma 7.4 *The lowest eigenvalue of the Dirichlet realization of the Laplacian in a relatively compact domain Ω is positive:*

$$\lambda_1 > 0. \tag{7.2.2}$$

Proof We know that $\lambda_1 \geq 0$; the Dirichlet realization of the Laplacian is positive. If $\lambda_1 = 0$, any corresponding normalized eigenfunction ϕ_1 satisfies

$$\langle -\Delta\phi_1,\, \phi_1 \rangle = 0$$

and, consequently, $\nabla\phi_1 = 0$ in Ω. This gives first the result that ϕ_1 is constant in each connected component of Ω, but because the trace of ϕ_1 on $\partial\Omega$ vanishes ($\phi_1 \in H_0^1(\Omega)$), this implies that $\phi_1 = 0$. So, we obtain a contradiction. $\qquad\square$

Finally, it follows from standard regularity theorems that the eigenfunctions belong (if Ω is regular) to $C^\infty(\overline{\Omega})$.

We now state the following proposition.

Proposition 7.5 *The lowest eigenvalue is simple, and the first eigenfunction can be chosen to be positive.*

The natural idea is to apply the Krein–Rutman theorem to $(-\Delta_D + 1)^{-1}$. One has to show that this operator is positivity-improving. This is indeed the case if the domain is connected, but the proof of this property will not be given here. We shall show only that $(-\Delta_D + I)^{-1}$ is positivity-preserving, and observe that this implies (following the proof of the Krein–Rutman theorem) that if Ω is an eigenfunction, then $|\Omega|$ is an eigenfunction. The proof of Proposition 7.5 will then be completed by using the properties of superharmonic functions.

Lemma 7.6 *The operator $(-\Delta_D + I)^{-1}$ is positivity-preserving.*

The proof is a consequence of the maximum principle. It is enough to show that

$$-\Delta u + u = f, \ \gamma_0 u = 0 \text{ and } f \geq 0 \text{ a.e.,}$$

implies that $u \geq 0$ almost everywhere.

We introduce $u^+ = \max(u, 0)$ and $u^- = -\inf(u, 0)$. A standard proposition (see below[3]) shows that u^+ and u^- belong to $H_0^1(\Omega)$. Multiplying by u^-, we obtain

$$\int \nabla u^+ \cdot \nabla u^- \, dx - \|\nabla u^-\|^2 - \|u^-\|^2 = \int_\Omega f u^- \, dx \geq 0,$$

which implies, using the corollary of the next proposition, that $u^- = 0$.

Proposition 7.7 *Suppose that $f \in L^1_{loc}(\mathbb{R}^m)$ with $\nabla f \in L^1_{loc}(\mathbb{R}^m)$. Then, also, $\nabla |f| \in L^1_{loc}(\mathbb{R}^m)$, and with the notation*

$$\text{sign } z = \begin{cases} \overline{z}/|z|, & z \neq 0, \\ 0, & z = 0, \end{cases} \tag{7.2.3}$$

---

[3] Recall that $|u| = u^+ + u^-$ and $u = u^+ - u^-$.

we have

$$\nabla |f|(x) = \text{Re}\{\text{sign}(f(x)) \nabla f(x)\} \text{ almost everywhere.} \tag{7.2.4}$$

In particular, we have $\big|\nabla |f|\big| \leq |\nabla f|$, *almost everywhere.*

Corollary 7.8 *Under the assumptions of Proposition 7.7 and if f is real,*

$$\nabla f^+(x) = \frac{1}{2}(\text{sign}(f(x)) + 1) \nabla f(x) \text{ almost everywhere} \tag{7.2.5}$$

and

$$\nabla f^-(x) = \frac{1}{2}(\text{sign}(f(x)) - 1) \nabla f(x) \text{ almost everywhere.} \tag{7.2.6}$$

Hence

$$\nabla f^+(x) \cdot \nabla f^-(x) = \frac{1}{4}(\text{sign}(f(x))^2 - 1) |\nabla f(x)|^2 \leq 0 \text{ almost everywhere.} \tag{7.2.7}$$

Proof of Proposition 7.7   Suppose first that $u \in C^\infty(\mathbb{R}^m)$, and define $|z|_\epsilon = \sqrt{|z|^2 + \epsilon^2} - \epsilon$, for $z \in \mathbb{C}$ and $\epsilon > 0$. We observe that

$$0 \leq |z|_\epsilon \leq |z| \text{ and } \lim_{\epsilon \to 0} |z|_\epsilon = |z|.$$

Then $|u|_\epsilon \in C^\infty(\mathbb{R}^m)$ and

$$\nabla |u|_\epsilon = \frac{\text{Re}\,(\bar{u}\,\nabla u)}{\sqrt{|u|^2 + \epsilon^2}}. \tag{7.2.8}$$

Now let f be as in the proposition, and define f_δ as the convolution

$$f_\delta = f * \rho_\delta,$$

where ρ_δ is a standard approximation of the unity for convolution. Explicitly, we take a $\rho \in C_0^\infty(\mathbb{R}^m)$ with

$$\rho \geq 0, \quad \int_{\mathbb{R}^m} \rho(x)\,dx = 1,$$

and define $\rho_\delta(x) := \delta^{-n}\rho(x/\delta)$, for $x \in \mathbb{R}^m$ and $\delta > 0$. Then $f_\delta \to f$, $|f_\delta| \to |f|$, and $\nabla f_\delta \to \nabla f$ in $L^1_{\text{loc}}(\mathbb{R}^m)$ as $\delta \to 0$.

Take a test function $\phi \in C_0^\infty(\mathbb{R}^m)$. We may extract a subsequence $\{\delta_k\}_{k\in\mathbb{N}}$ (with $\delta_k \to 0$ for $k \to \infty$) such that $f_{\delta_k}(x) \to f(x)$ for almost every $x \in \operatorname{supp}\phi$. We restrict our attention to this subsequence. For simplicity of notation, we shall omit the k from the notation and write $\lim_{\delta\to 0}$ instead of $\lim_{k\to\infty}$.

We now calculate, using dominated convergence and (7.2.8),

$$\int (\nabla\phi)\,|f|\,dx = \lim_{\epsilon\to 0} \int (\nabla\phi)\,|f|_\epsilon\,dx$$

$$= \lim_{\epsilon\to 0}\lim_{\delta\to 0} \int (\nabla\phi)\,|f_\delta|_\epsilon\,dx$$

$$= -\lim_{\epsilon\to 0}\lim_{\delta\to 0} \int \phi \frac{\operatorname{Re}(\overline{f_\delta}\nabla f_\delta)}{\sqrt{|f_\delta|^2 + \epsilon^2}}\,dx.$$

Using the pointwise convergence of $f_\delta(x)$ and $\|\nabla f_\delta - \nabla f\|_{L^1(\operatorname{supp}\phi)} \to 0$, we can take the limit $\delta \to 0$ and obtain

$$\int (\nabla\phi)\,|f|\,dx = -\lim_{\epsilon\to 0} \int \phi \frac{\operatorname{Re}(\overline{f}\nabla f)}{\sqrt{|f|^2 + \epsilon^2}}\,dx. \tag{7.2.9}$$

Now, $\phi\nabla f \in L^1(\mathbb{R}^m)$ and $\overline{f(x)}/\sqrt{|f|^2 + \epsilon^2} \to \operatorname{sign} f(x)$ as $\epsilon \to 0$, so we obtain (7.2.4) from (7.2.9) by dominated convergence. $\square$

We can now look directly at the property (which is a consequence of this positivity-improving property, but we have not given a proof) that the first eigenfunction does not vanish. This eigenfunction is positive in Ω, belongs to $C^\infty(\overline{\Omega})$ by a regularity theorem, and satisfies

$$-\Delta u = \lambda u \geq 0. \tag{7.2.10}$$

Hence u is superharmonic and satisfies the following mean value property: for all $y \in \Omega$, and for all $R > 0$ such that $B(y, R) \Subset \Omega$, we have

$$u(y) \geq \frac{1}{\operatorname{vol}(B(y, R))} \int_{B(y,R)} u(z)\,dz.$$

Moreover, we know that $\inf u = 0$. By applying this mean value property with y (if any) such that $u(y) = 0$, we obtain the result that $u = 0$ in $B(y, R)$. Using in addition a connectedness argument, we obtain the result that in a connected open set, u is either identically 0 or positive.

Let us now return to what appears in the proof of the Krein–Rutman theorem (Theorem 6.40). Here, u_λ^+ and u_λ^- either are 0 or are positive eigenfunctions, and we also have $\langle u_\lambda^+, u_\lambda^- \rangle = 0$. Hence u_λ^+ or u_λ^- should vanish. We can then show the simplicity of the first eigenvalue as in the proof of the Krein–Rutman theorem. Hence we have finally completed the proof of Proposition 7.5.

Remark 7.9 (Extension to operators with a compact resolvent.) What we have done for the spectral analysis of the Dirichlet realization is in fact quite general. It can be applied to self-adjoint operators that are bounded from below and have a compact resolvent. We show that in this case there exists an infinite sequence (if the Hilbert space is infinite dimensional) of real eigenvalues λ_n tending to $+\infty$ such that the corresponding eigenspaces are mutually orthogonal, of finite dimension, and such that their corresponding Hilbertian sum is equal to $\mathcal{H}$. Typically, the method is applied to the Neumann realization of the Laplacian in a relatively compact domain Ω in $\mathbb{R}^m$ and to the harmonic oscillator in $\mathbb{R}^m$. We shall describe various examples in the following sections.

7.3 Operators with a compact resolvent: the Schrödinger operator in an unbounded domain

In this section, we recall some criteria for compactness of the resolvent of the Schrödinger operator $P = -\Delta + V$ in $\mathbb{R}^m$ in connection with the precompactness criterion. In the case of the Schrödinger equation on $\mathbb{R}^m$ and if V is C^∞ and bounded from below by a constant $-C$, the domain of the self-adjoint extension is always contained in

$$Q(P) := H_V^1(\mathbb{R}^m) = \{u \in H^1(\mathbb{R}^m) | (V + C)^{1/2} u \in L^2(\mathbb{R}^m)\}.$$

$Q(P)$ is usually called the form domain of the form

$$u \mapsto \int_{\mathbb{R}^m} |\nabla u(x)|^2 \, dx + \int_{\mathbb{R}^m} V(x) \, |u(x)|^2 \, dx.$$

It is easy to see that if V tends to ∞, then the injection of H_V^1 in L^2 is compact (using a precompactness criterion). We then obtain, observing that $(P + 1)^{-1}$ is continuous from L^2 into H_V^1, the result that the resolvent $(P + \lambda)^{-1}$ is a compact operator for $\lambda \notin \sigma(P)$.

In the case of a compact manifold M and if we consider the Laplace–Beltrami operator on M, the compactness of the resolvent can be obtained without additional assumptions about V. The domain of the operator is $H^2(M)$, and we have compact injection from $H^2(M)$ into $L^2(M)$.

The condition that $V \to \infty$ as $|x| \to \infty$ is not a necessary condition. We can in fact replace it by the following weaker sufficient condition.

Proposition 7.10 *If the injection of $H_V^1(\mathbb{R}^m)$ into $L^2(\mathbb{R}^m)$ is compact, then P has a compact resolvent.*

More concretely, the way to verify this criterion is to show the existence of a continuous function $x \mapsto \rho(x)$ tending to ∞ as $|x| \to +\infty$ such that

$$H_V^1(\mathbb{R}^m) \subset L_\rho^2(\mathbb{R}^m). \tag{7.3.1}$$

Of course, the preceding case corresponds to $\rho = V$, but, as a typical example of this strategy, we shall show in Example 7.12 that the Schrödinger operator on $\mathbb{R}^2$, $-\Delta + x^2 y^2 + 1$, has a compact inverse. On the other hand, the criterion that $V \to +\infty$ as $|x| \to +\infty$ is not too far from optimality.

We can in fact prove the following lemma.

Lemma 7.11 *Suppose that $V \geq 0$ and that there exists $r > 0$ and a sequence σ_n such that $|\sigma_n| \to +\infty$ and such that*

$$\sup_{n\in\mathbb{N}} \sup_{x\in B(\sigma_n,r)} V(x) < +\infty. \tag{7.3.2}$$

Then $-\Delta + V + 1$ does not have a compact inverse.

Proof Consider the sequence $\phi_n(x) = \psi(x - \sigma_n)$, where ψ is a compactly supported function of L^2-norm 1 and with support in $B(0, r)$. We observe that the ϕ_n are an orthonormal sequence (after possibly extracting a subsequence to obtain the result that the supports of ϕ_n and $\phi_{n'}$ are disjoint for $n \neq n'$) that satisfies, for some constant C,

$$\|\phi_n\|_{H^2}^2 + \|V\phi_n\|_{L^2}^2 \leq C. \tag{7.3.3}$$

In particular, there exists C such that

$$\|(-\Delta + V + 1)\phi_n\|_{L^2} \leq C, \ \forall n \in \mathbb{N}.$$

But we cannot extract a strongly convergent sequence in L^2 from this sequence, because ϕ_n is weakly convergent to 0 and $\|\phi_n\| = 1$. So, the operator $(-\Delta + V + I)^{-1}$ cannot be a compact operator. $\square$

Example 7.12 The unbounded operator on $L^2(\mathbb{R}^2)$

$$P := -\frac{d^2}{dx^2} - \frac{d^2}{dy^2} + x^2 y^2$$

has a compact resolvent.

Hint   We can introduce

$$X_1 = \frac{1}{i}\partial_x, \ X_2 = \frac{1}{i}\partial_y, \ X_3 = xy$$

and show, for $j = 1, 2$ and for a suitable constant C, the following inequality:

$$||(x^2 + y^2 + 1)^{-1/4}[X_j, X_3]u||^2 \leq C\left(\langle Pu, u\rangle_{L^2(\mathbb{R}^2)} + ||u||^2\right),$$

for all $u \in C_0^\infty(\mathbb{R}^2)$. We can also observe that

$$i[X_1, X_3] = y, \ i[X_2, X_3] = x,$$

and then

$$||(x^2 + y^2 + 1)^{-1/4}[X_1, X_3]u||^2 + ||(x^2 + y^2 + 1)^{-1/4}[X_2, X_3]u||^2 + ||u||^2$$
$$\geq ||(x^2 + y^2 + 1)^{1/4}u||^2.$$

To check $||(x^2 + y^2 + 1)^{-1/4}[X_1, X_3]u||^2$, we note that

$$||(x^2 + y^2 + 1)^{-1/4}[X_1, X_3]u||^2 = \left\langle \frac{-iy}{(x^2 + y^2 + 1)^{1/4}}u, \ (X_1 X_3 - X_3 X_1)u\right\rangle,$$

perform an integration by parts, and check a commutator.

7.4 The Schrödinger operator with a magnetic field

We consider on $\mathbb{R}^m$ the so-called Schrödinger operator with a magnetic field,

$$P_{A,V} := -\Delta_A + V, \tag{7.4.1}$$

where

$$-\Delta_A := \sum_{j=1}^m \left(\frac{1}{i}\partial_{x_j} - A_j(x)\right)^2. \tag{7.4.2}$$

Here $x \mapsto \vec{A} = (A_1(x), \cdots, A_n(x))$ is a vector field on $\mathbb{R}^m$, called the magnetic potential, and V is called the electric potential. It is easy to see that when V is semibounded, the operator is symmetric and semibounded on $C_0^\infty(\mathbb{R}^m)$. We can therefore consider the Friedrichs extension and analyze the properties of this self-adjoint extension.

A general question arises as to whether one can obtain an operator of the type $P_{A,V}$ that has a compact resolvent if $V = 0$. This problem is called the problem of the magnetic bottle.

A heuristic description is that the modulus of the magnetic field can in some sense play the role of an electric potential if it does not oscillate too rapidly ($m \geq 2$). To define the magnetic field, it is probably easier to consider the magnetic potential as a one-form

$$\sigma_A = \sum_{j=1}^{n} A_j(x)\, dx_j.$$

The magnetic field is then defined as the two-form

$$\omega_B = d\sigma_A = \sum_{j<k} (\partial_{x_j} A_k - \partial_{x_k} A_j)\, dx_j \wedge dx_k.$$

The case where $m = 2$ is particularly simple. In this case,

$$\omega_B = B\, dx_1 \wedge dx_2,$$

and we can identify ω_B with the function $x \mapsto B(x) = \mathrm{curl}\,(\vec{A})(x)$.

The proof is particularly simple in the case where $B(x)$ has a constant sign (say, $B(x) \geq 0$). In this case, we immediately have the inequality

$$\int_{\mathbb{R}^2} B(x)\, |u(x)|^2\, dx_1\, dx_2 \leq \langle -\Delta_A u, u \rangle_{L^2}. \tag{7.4.3}$$

We can observe the following identities between operators:

$$B(x) = \frac{1}{i}[X_1, X_2], \quad -\Delta_A = X_1^2 + X_2^2. \tag{7.4.4}$$

Here,

$$X_1 = \frac{1}{i}\partial_{x_1} - A_1(x), \quad X_2 = \frac{1}{i}\partial_{x_2} - A_2(x).$$

Note also that

$$\langle -\Delta_A u, u \rangle = ||X_1 u||^2 + ||X_2 u||^2,$$

for all $u \in C_0^\infty(\mathbb{R}^2)$. It is then easy to obtain (7.4.3) by an integration by parts. By introducing $Z = X_1 + iX_2$, one can also use the nonnegativity of Z^*Z or ZZ^*. We then easily obtain, as in the previous example, the result that the operator has a compact resolvent if $B(x) \to +\infty$.

As a simple example, we can think of

$$\vec{A} = \left(-x_1^2 x_2, +x_1 x_2^2\right),$$

which gives

$$B(x) = x_1^2 + x_2^2.$$

Note that the case $m = 2$ is rather particular and that it is more difficult to treat $m > 2$. We have in fact to introduce a partition of unity.

Note also that it is also not necessary that the magnetic field tends to $+\infty$. The magnetic Laplacian associated with

$$\vec{A} = \left(x_1^2 x_2, x_1 x_2^2\right),$$

which gives

$$B(x) = x_2^2 - x_1^2,$$

also has a compact resolvent.

7.5 Laplace–Beltrami operators on a Riemannian compact manifold

7.5.1 Generalities

If M is a compact Riemannian manifold, it is well known that in this case we can canonically define a measure $d\mu_M$ on M and, consequently, the Hilbertian space $L^2(M)$. We also have a canonical definition of the gradient. At each point x of M, we have a scalar product on $T_x M$, giving an isomorphism between the tangent space at x, $T_x M$, and its dual space, $T_x^* M$. Using this family of isomorphisms, we have a natural identification between the C^∞-vector fields on M and the 1-forms on M. In this identification, the vector field grad u associated with a C^∞ function on M corresponds to the 1-form du.

We consider on $C^\infty(M) \times C^\infty(M)$ the sesquilinear form

$$(u, v) \mapsto a_0(u, v) := \int_M \langle \operatorname{grad} u(x), \operatorname{grad} v(x) \rangle_{T_x M} \, d\mu_M.$$

There is a natural differential operator $-\Delta_M$, called the Laplace–Beltrami operator on M, such that

$$a_0(u, v) = \langle -\Delta_M u, v \rangle_{L^2(M)}.$$

In this context, it is not difficult to define the Friedrichs extension and to obtain a self-adjoint extension of $-\Delta_M$ as a self-adjoint operator on $L^2(M)$.

The domain is easily characterized as being $H^2(M)$, the Sobolev space naturally associated with $L^2(M)$, and it can be shown that the injection of $H^2(M)$ into $L^2(M)$ is compact because M is compact. The self-adjoint extension of $-\Delta_M$ has a compact resolvent, and the general theory can be applied to this example.

7.5.2 The case on the circle $\mathbb{S}^1$

The simplest model is the operator $-d^2/d\theta^2$ on the circle of radius one, whose spectrum is $\{n^2, n \in \mathbb{N}\}$. For $n > 0$, the multiplicity is 2. An orthonormal basis is given by the functions $\theta \mapsto (2\pi)^{-1/2} \exp in\theta$ for $n \in \mathbb{Z}$. Here, the form domain of the operator is $H^{1,\mathrm{per}}(S^1)$, and the domain of the operator is $H^{2,\mathrm{per}}(S^1)$. These spaces have two descriptions. One way is to describe these spaces by

$$H^{1,\mathrm{per}} := \{u \in H^1(]0, 2\pi[\mid u(0) = u(2\pi)\}$$

and

$$H^{2,\mathrm{per}} := \{u \in H^2(]0, 2\pi[\mid u(0) = u(2\pi), \, u'(0) = u'(2\pi)\}.$$

The other way is to consider the Fourier coefficients of u. The Fourier coefficients of $u \in H^{k,\mathrm{per}}$ are in $\mathbf{h}^k$. Here,

$$\mathbf{h}^k := \{u_n \in \ell^2(\mathbb{Z}) \mid n^k u_n \in \ell^2(\mathbb{Z})\}.$$

It is then easy to prove the compact injection from $H^{1,\mathrm{per}}$ in $L^2(S^1)$ or, equivalently, from $\mathbf{h}^1$ into ℓ^2.

More generally, it is known that elliptic symmetric nonnegative operators of order $m > 0$ admit a self-adjoint extension with a compact resolvent.

7.5.3 The Laplacian on $\mathbb{S}^2$

We can also consider the Laplacian on $\mathbb{S}^2$. We describe $\mathbb{S}^2$ by spherical coordinates as usual, with

$$x = \cos \phi \sin \theta, y = \sin \phi \sin \theta, z = \cos \theta, \text{ with } \phi \in [-\pi, \pi[\, , \, \theta \in]0, \pi[\, ,$$

(7.5.1)

and we add the "north" and "south" poles, corresponding to the two points $(0, 0, 1)$ and $(0, 0, -1)$. We look for eigenfunctions of the Friedrichs extension of

$$\mathbf{L}^2 = -\frac{1}{\sin^2 \theta} \frac{\partial^2}{\partial \phi^2} - \frac{1}{\sin \theta} \frac{\partial}{\partial \theta} \sin \theta \frac{\partial}{\partial \theta}$$

(7.5.2)

in $L^2(\sin\theta\,d\theta\,d\phi)$, satisfying

$$\mathbf{L}^2 Y_{\ell m} = \ell(\ell+1)Y_{\ell m}. \tag{7.5.3}$$

The standard spherical harmonics, corresponding to $\ell \geq 0$ and for an integer $m \in \{-\ell, \ldots, \ell\}$, are defined by

$$Y_{\ell m}(\theta, \phi) = c_{\ell,m}\exp im\phi\frac{1}{\sin^m\theta}\left(-\frac{1}{\sin\theta}\frac{d}{d\theta}\right)^{\ell-m}\sin^{2\ell}\theta, \tag{7.5.4}$$

where $c_{\ell,m}$ is an explicit normalization constant.

For future extensions, we prefer to take this as a definition for $m \geq 0$ and then to observe that

$$Y_{\ell,-m} = \hat{c}_{\ell,m}\overline{Y_{\ell,m}}. \tag{7.5.5}$$

For $\ell = 0$, we obtain $m = 0$ and the constant. For $\ell = 1$, we obtain for $m = 1$ the function $(\theta, \phi) \mapsto \sin\theta \exp i\phi$, for $m = -1$ the function $\sin\theta \exp{-i\phi}$, and for $m = 0$ the function $\cos\theta$, which shows that the multiplicity is 3 for the eigenvalue 2.

In order to show the completeness directly, it is enough to show that for a given $m \geq 0$, the orthogonal family (indexed by $\ell \in \{m + \mathbb{N}\}$) consisting of the functions

$$\theta \mapsto \psi_{\ell,m}(\theta) := \frac{1}{\sin^m\theta}\left(-\frac{1}{\sin\theta}\frac{d}{d\theta}\right)^{\ell-m}\sin^{2\ell}\theta$$

spans all $L^2(]0, \pi[, \sin\theta\,d\theta)$. For this, we consider $\chi \in C_0^\infty(]0, \pi[)$ and assume that

$$\int_0^\pi \chi(\theta)\psi_{\ell,m}(\theta)\sin\theta\,d\theta = 0, \ \forall\ell \in \{m + \mathbb{N}\}.$$

We would like to deduce that this implies $\chi = 0$. After a change of variable $t = \cos\theta$ and an integration by parts, this problem is equivalent to showing that if

$$\int_{-1}^1 \psi(t)((1-t^2)^\ell)^{(\ell-m)}(t)\,dt = 0, \ \forall\ell \in \{m + \mathbb{N}\},$$

then $\psi = 0$.

By observing that the space spanned by the functions

$$(1-t^2)^{-m}((1-t^2)^\ell)^{(\ell-m)}$$

(which are actually polynomials of exact order ℓ) is the space of all polynomials, we can conclude the completeness.

7.6 Notes

For the part of this chapter related to statistical mechanics, readers are referred to the author's book [He4] and references therein. The standard regularity theorems for elliptic operators that are used from Section 7.2 onward can be found in [Br] and [LiMa]. For Section 7.2, readers are referred to the book by Lieb and Loss [LiLo], for example. The material of Section 7.4 was produced in collaboration with A. Morame (see [He2] for a more complete presentation, or [HelN, FoHe] and references therein). This material involves a technique based on commutators initially developed by J. Kohn for the proof of the hypoellipticity of Hörmander's operators [Ho1], which we shall encounter again in Chapter 15. For the Laplace–Beltrami operator, readers are referred to the book by Berger, Gauduchon, and Mazet [BGM].

7.7 Exercises

Exercise 7.1 Let K be a kernel in $\mathcal{S}(\mathbb{R}^2)$ that is positive and symmetric.

(a) Show that the associated operator $\mathcal{K}$, which is defined on $\mathcal{S}(\mathbb{R})$ by

$$(\mathcal{K}u)(x) = \int_{\mathbb{R}} K(x, y)u(y)\, dy,$$

can be extended as a compact operator on $L^2(\mathbb{R})$.

(b) Let I be an open interval in $\mathbb{R}$, and denote by $\mathcal{K}_I$ the operator on $L^2(I)$ defined by

$$(\mathcal{K}_I u)(x) = \int_{I} K(x, y)u(y)\, dy.$$

Let λ_I^1 be the largest eigenvalue of $\mathcal{K}_I$. Show that

$$\lambda_I^1 \le \lambda_{\mathbb{R}}^1.$$

Show that we have strict inequality when I is not $\mathbb{R}$.

(c) Let u^1 be a normalized eigenfunction of $\mathcal{K}$ associated with $\lambda_{\mathbb{R}}^1$. Using its restriction to I, show the inequality

$$\lambda_{\mathbb{R}}^1 \le \lambda_I^1 \left(1 - \|u^1\|_{L^2(I^c)}\right)^{-1}.$$

(d) Let $I_n = [-n, n]$. Show that $\lambda_{I_n}^1$ converges rapidly to $\lambda_{\mathbb{R}}^1$ as $n \to +\infty$. More precisely, show that for all $j \in \mathbb{N}$, there exists a constant C_j such that

$$|\lambda_{\mathbb{R}}^1 - \lambda_{I_n}^1| \le C_j\, n^{-j}, \ \forall n \in \mathbb{N}^*.$$

Exercise 7.2 Show that the Friedrichs extension in $L^2(\mathbb{R}^2)$ of

$$T := -\left(\frac{d}{dx_1} - ix_2x_1^4\right)^2 - \frac{d^2}{dx_2^2} + x_2^4$$

has a compact resolvent.

Exercise 7.3 Consider in $\Omega =]0, 1[\times \mathbb{R}$ a positive C^∞ function V, and let S_0 be the Schrödinger operator $S_0 = -\Delta + V$ defined on $C_0^\infty(\Omega)$.

(a) Show that S_0 admits a Friedrichs self-adjoint extension on $L^2(\Omega)$. Let S be this extension.

(b) Determine if S has a compact resolvent in the following cases:

1. $V(x) = 0$.
2. $V(x) = x_1^2 + x_2^2$.
3. $V(x) = x_1^2$.
4. $V(x) = x_2^2$.
5. $V(x) = (x_1 - x_2)^2$.

Determine the spectrum in cases 1 and 4. We suggest that you first determine the spectrum of the Dirichlet realization (or the Neumann realization) of $-d^2/dx^2$ on $]0, 1[$.

Exercise 7.4 Let $\delta \in \mathbb{R}$.

(a) Show that the operator P_δ defined on $C_0^\infty(\mathbb{R}^2)$ by

$$P_\delta := D_x^2 + D_y^2 + x^2y^4 + x^4y^2 + \delta(x + y)$$

is semibounded. We suggest that you first show the inequality

$$\langle P_0u, u\rangle \geq ||xu||^2 + ||yu||^2.$$

(b) Show that there exists a natural self-adjoint extension of P_δ.
(c) What is the corresponding form domain?
(d) Show that the self-adjoint extension has a compact resolvent.

Exercise 7.5 Show that with the differential operator on $C_0^\infty(\mathbb{R} \times]0, 1[)$

$$T_0 := (D_{x_1} - x_2x_1^2)^2 + (D_{x_2})^2,$$

one can associate an unbounded self-adjoint operator T on $L^2(\mathbb{R} \times]0, 1[)$ with a compact resolvent.

Exercise 7.6 Show that $-\Delta_A$ has a compact resolvent (see Section 7.4), under the assumption that $|B(x)| \to +\infty$ as $|x| \to +\infty$ and that ∇B is bounded.

Exercise 7.7 (Witten Laplacian.) Let ϕ be a C^2 function on $\mathbb{R}^m$ such that $|\nabla\phi(x)| \to +\infty$ as $|x| \to +\infty$ and that ϕ has uniformly bounded second derivatives. Consider the differential operator on $C_0^\infty(\mathbb{R}^2)$

$$-\Delta + 2\nabla\phi \cdot \nabla.$$

We consider this operator as an unbounded operator on $\mathcal{H} = L^2(\mathbb{R}^m, \exp -2\phi \, dx)$.

(a) Show that it admits a self-adjoint extension and that its spectrum is discrete.

(b) Assume in addition that $\int_{\mathbb{R}^m} \exp -2\phi \, dx < +\infty$. Show that its lowest eigenvalue is simple, and determine a corresponding eigenvector.

8

Self-adjoint unbounded operators
and spectral theory

When the theory of compact operators or the theory of operators with a compact resolvent cannot be used, there is a need, in the self-adjoint case, for a theorem to replace the diagonalization theorem. The aim of this chapter is to present such a theorem for self-adjoint operators, called the spectral theorem; to give some elements of the proof, which involves nice functional analysis; and to show some rather simple but very useful applications of this theorem.

8.1 The theory of operators with a compact resolvent revisited

We assume that $\mathcal{H}$ is a Hilbert space. Before we describe the general case, let us return to the spectral theorem for compact operators T or operators with a compact resolvent. This will permit us to introduce a new vocabulary. We have seen that we can obtain a decomposition of $\mathcal{H}$ in the form

$$\mathcal{H} = \oplus_{k \in \mathbb{N}} V_k, \tag{8.1.1}$$

such that

$$T u_k = \lambda_k u_k, \text{ if } u_k \in V_k. \tag{8.1.2}$$

Hence we have decomposed $\mathcal{H}$ into a direct sum of orthogonal subspaces V_k in which the self-adjoint operator T is reduced to multiplication by λ_k.

If P_k denotes the orthogonal projection operator onto V_k, we can write

$$I = \sum_k P_k \tag{8.1.3}$$

(where the limit is in the strong-convergence sense), and

$$T u = \sum_k \lambda_k P_k u, \ \forall u \in D(T). \tag{8.1.4}$$

Here we recall the following definition.

Definition 8.1 An operator $P \in \mathcal{L}(\mathcal{H})$ is called an orthogonal projection if $P = P^*$ and $P^2 = P$.

If we assume that T is semibounded (with a compact resolvent[1]), we can introduce for any $\lambda \in \mathbb{R}$

$$G_\lambda = \oplus_{\lambda_k \leq \lambda} V_k, \tag{8.1.5}$$

and E_λ is the orthogonal projection onto G_λ:

$$E_\lambda = \sum_{\lambda_k \leq \lambda} P_k. \tag{8.1.6}$$

It is easy to see that the function $\lambda \mapsto E_\lambda$ has values in $\mathcal{L}(\mathcal{H})$ and that these satisfy the following properties:

- $E_\lambda = E_\lambda^*$;
- $E_\lambda \cdot E_\mu = E_{\inf(\lambda,\mu)}$;
- for all λ, $E_{\lambda+0} = E_\lambda$;
- $\lim_{\lambda \to -\infty} E_\lambda = 0$, $\lim_{\lambda \to +\infty} E_\lambda = Id$.
- $E_\lambda \geq 0$.

All of the above limits are in the sense of strong convergence.

We also observe that

$$E_{\lambda_k} - E_{\lambda_k - 0} = P_k.$$

Then, in the sense of vector-valued distributions, we have

$$dE_\lambda = \sum_k \delta_{\lambda_k} \otimes P_k, \tag{8.1.7}$$

where δ_{λ_k} is the Dirac measure at the point λ_k. Hence, in the sense of Stieltjes integrals (this will be explained in more detail below), we can write

$$x = \int_{-\infty}^{+\infty} dE_\lambda(x)$$

and

$$T = \int_{-\infty}^{+\infty} \lambda \, dE_\lambda.$$

This is the form to which we shall generalize the previous formulas in the case of any self-adjoint operator T.

---

[1] Note that if $(T - \lambda_0)^{-1}$ is compact for some $\lambda_0 \in \varrho(T)$, then this is true for any $\lambda \in \rho(T)$.

Functional calculus for operators with a compact resolvent If f is a continuous (or piecewise continuous) function, we can also define $f(T)$ as

$$f(T) = \sum_k f(\lambda_k) \cdot P_k,$$

which is an unbounded operator whose domain is

$$D(f(T)) = \{x \in \mathcal{H} \mid \sum_k |f(\lambda_k)|^2 \, \|x_k\|^2 < +\infty\},$$

where $x_k = P_k x$. We can also write $f(T)$ in the form

$$\langle f(T)x, \, y \rangle_{\mathcal{H}} = \int_{\mathbb{R}} f(\lambda) \, d\langle E_\lambda x, \, y \rangle,$$

where the domain of $f(T)$ is described as

$$D(f(T)) = \{x \in \mathcal{H} \mid \int_{\mathbb{R}} |f(\lambda)|^2 \, d\langle E_\lambda x, \, x \rangle_{\mathcal{H}} < +\infty\}.$$

Remark 8.2 For semibounded operators with a compact resolvent, there are two possible conventions for the notation for the eigenvalues. The first convention is to classify them into an increasing sequence

$$\mu_j \leq \mu_{j+1},$$

counting each eigenvalue according to its multiplicity. The second is to describe them as an increasing sequence λ_k where each λ_k is an eigenvalue of multiplicity m_k.

We now present a list of properties that are easy to verify in this particular case and will still be true in the general case.

1. If f and g coincide on $\sigma(T)$, then $f(T) = g(T)$. For any $(x, y) \in \mathcal{H} \times \mathcal{H}$, the support of the measure $d \langle E_\lambda x, \, y \rangle$ is contained[2] in $\sigma(T)$.
2. If f and g are functions on $\mathbb{R}$, then

$$f(T)g(T) = (f \cdot g)(T).$$

In particular, if $(T - z)$ is invertible, the inverse is given by $f(T)$, where f is a continuous function such that $f(\lambda) = (\lambda - z)^{-1}$, $\forall \lambda \in \sigma(T)$.

---

[2] We have not yet defined $\sigma(T)$ when T is unbounded. Think, in the case of operators with a compact resolvent, of the set of the eigenvalues!

3. If f is bounded, then $f(T)$ is bounded and we have

$$\|f(T)\| \leq \sup_{\lambda \in \sigma(T)} |f(\lambda)|. \tag{8.1.8}$$

4. The function f may be complex. Note that in this case we obtain

$$f(T)^\star = \overline{f}(T). \tag{8.1.9}$$

An interesting case, for $z \in \mathbb{C} \setminus \mathbb{R}$, is the function $\lambda \mapsto (\lambda - z)^{-1}$. In this case, we obtain from (8.1.8)

$$\|(T - z)^{-1}\|_{\mathcal{L}(\mathcal{H})} \leq |\operatorname{Im} z|^{-1}. \tag{8.1.10}$$

5. More generally, this also works for $z \in \mathbb{R} \setminus \sigma(T)$. We obtain in this case the spectral theorem

$$\|(T - z)^{-1}\|_{\mathcal{L}(\mathcal{H})} \leq d(z, \sigma(T))^{-1}. \tag{8.1.11}$$

6. If $f \in C_0^\infty(\mathbb{R})$, we have

$$f(T) = \frac{1}{\pi} \lim_{\epsilon \to 0^+} \int_{|\operatorname{Im} z| > \epsilon} \left(\frac{\partial \tilde{f}}{\partial \bar{z}} \right) (T - z)^{-1} \, dx \, dy. \tag{8.1.12}$$

Here, $\tilde{f}$ is defined by

$$\tilde{f}(x, y) = (f(x) + iyf'(x))\chi(y),$$

where χ is equal to 1 in a neighborhood of 0 and has compact support. This formula can be proven using the Green–Riemann formula (prove it first with T replaced by the scalar λ) or using the fact that

$$\partial_{\bar{z}} \frac{1}{z - \lambda} = \pi \delta_{(\lambda,0)},$$

where $\delta_{(\lambda,0)}$ is the Dirac measure at $(\lambda, 0) \in \mathbb{R}^2$ and $\partial_{\bar{z}} = \frac{1}{2}(\partial_x + i\partial_y)$. One should observe that $\tilde{f}$ is not holomorphic, but it is an almost analytic extension of f in the sense that

$$\partial \tilde{f}/\partial \bar{z} = \mathcal{O}(y),$$

as $y \to 0$.

Remark 8.3 More generally, it can be shown that for any $N \geq 1$, there exists an almost analytic extension $\tilde{f} = \tilde{f}^N$ of f such that in addition

$$\partial \tilde{f} / \partial \bar{z} = \mathcal{O}(y^N),$$

as $y \to 0$.

8.2 The spectrum

We now return to the notion of a spectrum, which we have previously encountered only for bounded operators.

Definition 8.4 The resolvent set of a closed operator T is the set of the λ in $\mathbb{C}$ such that the range of $(T - \lambda)$ is equal to $\mathcal{H}$ and such that $(T - \lambda)$ admits a continuous operator, denoted by $R(\lambda)$, whose range is included in $D(T)$ such that

$$R(\lambda)(T - \lambda) = I_{D(T)}$$

and

$$(T - \lambda)R(\lambda) = I_{\mathcal{H}}.$$

As in the bounded case, we observe that the resolvent set is open (see the argument around (6.3.2)). Note also that the continuity of $R(\lambda)$ is actually a consequence of the property that the graph of $R(\lambda)$ is closed (using the fact that T is closed) and that $R(\lambda)$ is defined on $\mathcal{H}$.

Definition 8.5 The spectrum of a closed operator T is defined as the complementary set in $\mathbb{C}$ of the resolvent set.

It is then rather easy to show that the spectrum $\sigma(T)$ is closed in $\mathbb{C}$.

Lemma 8.6 *If T is self-adjoint, the spectrum $\sigma(T)$ is in $\mathbb{R}$ and not empty.*

The proof of the fact that the spectrum is contained in $\mathbb{R}$ if T is self-adjoint is very close to the bounded case.

The proof of nonemptiness is by contradiction. If T has an empty spectrum, T^{-1} is a bounded self-adjoint operator with a spectrum contained in $\{0\}$. We observe in fact that for $\lambda \neq 0$, the inverse of $T^{-1} - \lambda$ is given, if $\lambda^{-1} \in \varrho(T)$, by $\lambda^{-1} T(T - \lambda^{-1})^{-1}$. Hence, by our analysis of the spectrum of a bounded self-adjoint operator, $T^{-1} = 0$, which contradicts $T \circ T^{-1} = I$. This is no longer true in the non-self-adjoint case. In Chapter 14, we shall give an example of a non-self-adjoint operator with an empty spectrum that appears naturally in various questions in fluid mechanics.

8.3 Spectral family and resolution of the identity

Definition 8.7 A family of orthogonal projectors $E(\lambda)$ (also denoted by E_λ), $-\infty < \lambda < \infty$, in a Hilbert space $\mathcal{H}$ is called a resolution of the identity (or spectral family) if it satisfies the following conditions:

-
$$E(\lambda)E(\mu) = E(\min(\lambda, \mu)); \tag{8.3.1}$$

-
$$E(-\infty) = 0, \ E(+\infty) = I, \tag{8.3.2}$$

where $E(\pm\infty)$ is defined[3] by

$$E(\pm\infty)x = \lim_{\lambda \to \pm\infty} E(\lambda)x \tag{8.3.3}$$

for all x in $\mathcal{H}$;

-
$$E(\lambda + 0) = E(\lambda), \tag{8.3.4}$$

where $E(\lambda + 0)$ is defined by

$$E(\lambda + 0)x = \lim_{\mu \to \lambda, \ \mu > \lambda} E(\mu)x. \tag{8.3.5}$$

We have given an example of such a family in Section 8.1.

Proposition 8.8 *Let $E(\lambda)$ be a resolution of the identity (= spectral family); then, for all $x, y \in \mathcal{H}$, the function*

$$\lambda \mapsto \langle E(\lambda)x, \ y \rangle \tag{8.3.6}$$

is a function of bounded variation whose total variation[4] satisfies

$$V(x, y) \le ||x|| \cdot ||y||, \ \forall x, y \in \mathcal{H}. \tag{8.3.7}$$

Proof Let $\lambda_1 < \lambda_2 < \cdots < \lambda_n$. We first obtain, from the assumption (8.3.1), the result that

$$E_{]\alpha,\beta]} = E_\beta - E_\alpha$$

---

[3] Equation (8.3.1) gives the existence of the limit (see also Lemma 8.9). The limit in (8.3.3) is taken in $\mathcal{H}$. We observe that in fact $\lambda \mapsto \langle E(\lambda)x, \ x \rangle = ||E(\lambda)x||^2$ is monotonically increasing.

[4] See the definition in (8.3.9).

is an orthogonal projection. From the Cauchy–Schwarz inequality, we have

$$
\begin{aligned}
\sum_{j=2}^{n} |\langle E_{]\lambda_{j-1},\lambda_j]}x, y\rangle| &= \sum_{j=2}^{n} |\langle E_{]\lambda_{j-1},\lambda_j]}x, E_{]\lambda_{j-1},\lambda_j]}y\rangle| \\
&\leq \sum_{j=2}^{n} \|E_{]\lambda_{j-1},\lambda_j]}x\| \, \|E_{]\lambda_{j-1},\lambda_j]}y\|\| \\
&\leq \left(\sum_{j=2}^{n} \|E_{]\lambda_{j-1},\lambda_j]}x\|^2 \right)^{1/2} \left(\sum_{j=2}^{n} \|E_{]\lambda_{j-1},\lambda_j]}y\|^2 \right)^{1/2} \\
&= \left(\|E_{]\lambda_1,\lambda_n]}x\|^2 \right)^{1/2} \left(\|E_{]\lambda_1,\lambda_n]}y\|^2 \right)^{1/2}.
\end{aligned}
$$

But for $m > n$, we obtain

$$
\|x\|^2 \geq \|E_{]\lambda_n,\lambda_m]}x\|^2 = \sum_{i=n}^{m-1} \|E_{]\lambda_i,\lambda_{i+1}]}x\|^2. \tag{8.3.8}
$$

We finally obtain the result that for any finite sequence $\lambda_1 < \lambda_2 < \cdots < \lambda_n$, we have

$$
\sum_{j=2}^{n} |\langle E_{]\lambda_{j-1},\lambda_j]}x, y\rangle| \leq \|x\| \cdot \|y\|.
$$

This shows the bounded variation and the estimate of the total variation, defined as

$$
V(x, y) := \sup_{\lambda_1,\cdots,\lambda_n} \sum_{j=2}^{n} |\langle E_{]\lambda_{j-1},\lambda_j]}x, y\rangle|. \tag{8.3.9}
$$

$\square$

Hence we have shown that for all x and y in $\mathcal{H}$, the function $\lambda \mapsto \langle E(\lambda)x, y\rangle$ has bounded variation, and we can then show the existence of $E(\lambda + 0)$ and $E(\lambda - 0)$. This is the subject of the following lemma.

Lemma 8.9 *If $E(\lambda)$ is a family of projectors satisfying (8.3.1) and (8.3.2), then, for all $\lambda \in \mathbb{R}$, the operators*

$$
E_{\lambda+0} = \lim_{\mu \to \lambda \ \mu > \lambda} E(\mu), \quad E_{\lambda-0} = \lim_{\mu \to \lambda \ \mu < \lambda} E(\mu) \tag{8.3.10}
$$

are well defined when the limit is considered in the strong-convergence topology.

Proof Let us show the existence of the left limit. From (8.3.8), we obtain the result that for any $\epsilon > 0$, there exists $\lambda_0 < \lambda$ such that, $\forall \lambda'$, $\forall \lambda'' \in [\lambda_0, \lambda[$ such that $\lambda' < \lambda''$,

$$||E_{]\lambda', \lambda'']}x||^2 \leq \epsilon.$$

It is then easy to show that $E_{\lambda - 1/n}x$ is a Cauchy sequence converging to a limit and that this limit does not depend on the choice of the sequence tending to λ.

The proof of the existence of the limit from the right is the same. This ends the proof of the lemma. $\qquad\square$

It is then classical (Stieltjes integrals) that one can define, for any continuous complex-valued function $\lambda \mapsto f(\lambda)$, the integrals $\int_a^b f(\lambda) \, d\langle E(\lambda)x, y\rangle$ as a limit[5] of Riemann sums.

Proposition 8.10 *Let f be a continuous function on $\mathbb{R}$ with complex values and let $x \in \mathcal{H}$. Then it is possible to define, for $\alpha < \beta$, the integral*

$$\int_\alpha^\beta f(\lambda) \, dE_\lambda x$$

as the strong limit in $\mathcal{H}$ of the Riemann sum

$$\sum_j f(\lambda'_j)(E_{\lambda_{j+1}} - E_{\lambda_j})x, \qquad (8.3.11)$$

where

$$\alpha = \lambda_1 < \lambda_2 < \cdots < \lambda_n = \beta$$

and

$$\lambda'_j \in]\lambda_j, \lambda_{j+1}],$$

when $\max_j |\lambda_{j+1} - \lambda_j| \to 0$.

The proof is easy using the uniform continuity of f. Note also that the notation might be misleading. Writing $\int_{]\alpha,\beta]} f(\lambda) \, dE_\lambda x$ is less ambiguous.

We now arrive, as in the standard theory, at the generalized integral.

Definition 8.11 For any given $x \in \mathcal{H}$ and any continuous function f on $\mathbb{R}$, the integral

$$\int_{-\infty}^{+\infty} f(\lambda) \, dE_\lambda x$$

---

[5] The best way is to first consider the case where $x = y$ and then use a depolarization formula, in the same way that when we have a Hilbertian norm, we can recover the scalar product from the norm.

is defined as the strong limit in $\mathcal{H}$, if it exists, of $\int_\alpha^\beta f(\lambda)\,dE_\lambda x$ when $\alpha \to -\infty$ and $\beta \to +\infty$.

We shall assume that the theory works more generally for any Borelian function. This extension can be important, because we are interested in particular in the case where $f(t) = 1_{]-\infty,\lambda]}(t)$. More precisely, we have the following statement.

Theorem 8.12

1. *If μ is a complex Borel measure on $\mathbb{R}$ and if*

$$(\star) \quad f(x) = \mu(]-\infty, x]), \ \forall x \in \mathbb{R},$$

then f is a normalized function with bounded variation (NBV). By NBV, we mean with bounded variation, but also continuous from the right and such that $\lim_{x \to -\infty} f(x) = 0$.
2. *Conversely, to every $f \in NBV$, there corresponds a unique complex Borel measure μ such that $(\star)$ is satisfied.*

Theorem 8.13 *For a given x in $\mathcal{H}$ and if f is a complex-valued continuous function on $\mathbb{R}$, the following conditions are equivalent:*

- $$\int_{-\infty}^{+\infty} f(\lambda)\,dE_\lambda x \text{ exists}; \tag{8.3.12}$$

- $$\int_{-\infty}^{+\infty} |f(\lambda)|^2\,d\|E_\lambda x\|^2 < +\infty; \tag{8.3.13}$$

- $$y \mapsto \int_{-\infty}^{+\infty} f(\lambda)\,d(\langle E_\lambda y, x\rangle_{\mathcal{H}}) \tag{8.3.14}$$

is a continuous linear form.

Hint for the proof

(a) (8.3.12) implies (8.3.14), essentially by using repeatedly the Banach–Steinhaus theorem (also called the uniform boundedness theorem) and the definition of the integral.
(b) Let us prove that (8.3.14) implies (8.3.13).
   Let F be the linear form appearing in (8.3.14). If we introduce

$$y = \int_\alpha^\beta \overline{f(\lambda)}\,dE_\lambda x,$$

then we first observe (returning to Riemann integrals) that

$$y = E_{]\alpha,\beta]}y.$$

It is then not too difficult to show that

$$\overline{F(y)} = \int_{-\infty}^{+\infty} \overline{f(\lambda)} \, d\langle E_\lambda x, y\rangle$$

$$= \int_{-\infty}^{+\infty} \overline{f(\lambda)} \, d\langle E_\lambda x, E_{]\alpha,\beta]}y\rangle$$

$$= \int_{-\infty}^{+\infty} \overline{f(\lambda)} \, d\langle E_{]\alpha,\beta]}E_\lambda x, y\rangle$$

$$= \int_{\alpha}^{\beta} \overline{f(\lambda)} \, d\langle E_\lambda x, y\rangle$$

$$= ||y||^2.$$

Using (8.3.14), we obtain $||y||^2 \leq ||F|| \cdot ||y||$ and, consequently,

$$||y|| \leq ||F||. \tag{8.3.15}$$

Here we observe that the right-hand side is independent of α and β.

On the other hand, returning to Riemann sums, we obtain

$$||y||^2 = \int_{\alpha}^{\beta} |f(\lambda)|^2 \, d||E_\lambda x||^2.$$

We finally obtain

$$\int_{\alpha}^{\beta} |f(\lambda)|^2 \, d||E_\lambda x||^2 \leq ||F||^2. \tag{8.3.16}$$

Hence, taking the limits $\alpha \to -\infty$ and $\beta \to +\infty$, we obtain (8.3.13).

(c) For the last implication, it is enough to observe that for $\alpha' < \alpha < \beta < \beta'$, we have

$$\left\| \int_{\alpha'}^{\beta'} f(\lambda) \, dE_\lambda x - \int_{\alpha}^{\beta} f(\lambda) \, dE_\lambda x \right\|^2$$

$$= \int_{\alpha'}^{\alpha} |f(\lambda)|^2 \, d||E_\lambda x||^2 + \int_{\beta}^{\beta'} |f(\lambda)|^2 \, d||E_\lambda x||^2.$$

Theorem 8.14 *Let $\lambda \mapsto f(\lambda)$ be a real-valued continuous function. Let*

$$D_f := \{x \in \mathcal{H}, \int_{-\infty}^{+\infty} |f(\lambda)|^2 \, d\langle E_\lambda x, x\rangle < \infty\}.$$

Then D_f is dense in $\mathcal{H}$, and we can define T_f, whose domain is

$$D(T_f) = D_f,$$

and

$$\langle T_f x, \, y \rangle = \int_{-\infty}^{+\infty} f(\lambda) \, d\langle E_\lambda x, \, y \rangle$$

for all x in $D(T_f)$ and y in $\mathcal{H}$. The operator T_f is self-adjoint, and $T_f E_\lambda$ is an extension of $E_\lambda T_f$.

Proof of Theorem 8.14 From Property (8.3.2), we obtain the result that for any y in $\mathcal{H}$, there exists a sequence (α_n, β_n) such that $E_{]\alpha_n, \beta_n]} y \rightarrow y$ as $n \rightarrow +\infty$. But $E_{]\alpha, \beta]} y$ belongs to D_f, for any α, β, and this shows the density of D_f in $\mathcal{H}$.

We now observe that since f is real and E_λ is symmetric, the symmetry is clear. The self-adjointness is proven by using Theorem 8.13.   □

We observe that for $f_0 = 1$, we have $T_{f_0} = I$ and for $f_1(\lambda) = \lambda$, we obtain a self-adjoint operator $T_{f_1} := T$. In this case we say that

$$T = \int_{-\infty}^{+\infty} \lambda \, dE_\lambda$$

is a spectral decomposition of T, and we note that

$$\|Tx\|^2 = \int_{-\infty}^{+\infty} \lambda^2 \, d\langle E_\lambda x, \, x \rangle = \int_{-\infty}^{+\infty} \lambda^2 \, d\|E_\lambda x\|^2$$

for $x \in D(T)$. More generally,

$$\|T_f x\|^2 = \int_{-\infty}^{+\infty} |f(\lambda)|^2 \, d(\langle E_\lambda x, \, x \rangle) = \int_{-\infty}^{+\infty} |f(\lambda)|^2 \, d(\|E_\lambda x\|^2)$$

for $x \in D(T_f)$.

Conclusion   We have seen in this section how one can associate a self-adjoint operator with a spectral family of projectors. We have seen in Section 8.1 that the converse is true for a compact operator and for an operator with a compact resolvent. It remains to prove that this is true in the general case.

8.4 The spectral decomposition theorem

The spectral decomposition theorem makes it explicit that the preceding situation is actually the general one.

Theorem 8.15 *Any self-adjoint operator T in a Hilbert space $\mathcal{H}$ admits a spectral decomposition such that*

$$\langle Tx, y \rangle = \int_{\mathbb{R}} \lambda \, d\langle E_\lambda x, y \rangle_{\mathcal{H}},$$
$$Tx = \int_{\mathbb{R}} \lambda \, d(E_\lambda x). \tag{8.4.1}$$

Proof We shall give only the main points of the proof.

Step 1. It is natural to imagine that it is essentially enough to treat the case where T is a bounded self-adjoint operator (or at least a normal bounded operator, that is, an operator satisfying $T^*T = TT^*$). If A is instead a general semibounded self-adjoint operator, we can return to the bounded case by considering $(A + \lambda_0)^{-1}$, with λ_0 real, which is bounded and self-adjoint. In the general case,[6] we can consider $(A + i)^{-1}$.

Step 2. We analyze first the spectrum of $P(T)$, where P is a polynomial.

Lemma 8.16 *If P is a polynomial, then*

$$\sigma(P(T)) = \{P(\lambda) \mid \lambda \in \sigma(T)\}. \tag{8.4.2}$$

Proof We start from the identity $P(x) - P(\lambda) = (x - \lambda)Q_\lambda(x)$ and the corresponding identity between bounded operators $P(T) - P(\lambda) = (T - \lambda)Q_\lambda(T)$. This permits us to construct the inverse of $(T - \lambda)$ if we know the inverse of $P(T) - P(\lambda)$.

Conversely, we observe that if $z \in \mathbb{C}$ and if $\lambda_j(z)$ are the roots of $\lambda \mapsto (P(\lambda) - z)$, then we can write, for some $c \neq 0$,

$$(P(T) - z) = c \prod_j (T - \lambda_j(z)).$$

This permits us to construct the inverse of $(P(T) - z)$ if we have the inverses of $(T - \lambda_j(z))$ (for all j). $\qquad\square$

---

[6] Here we recall that an example of an operator that is not semibounded was given in Exercise 4.1.

Lemma 8.17 *Let T be a bounded self-adjoint operator. Then*

$$\|P(T)\| = \sup_{\lambda \in \sigma(T)} |P(\lambda)|. \tag{8.4.3}$$

We first observe that

$$\|P(T)\|^2 = \|P(T)^* P(T)\|.$$

This is a consequence of the general property of bounded linear operators that

$$\|A^* A\| = \|A\|^2. \tag{8.4.4}$$

We recall that the proof is obtained by observing first that

$$
\begin{aligned}
\|A^* A\| &= \sup_{\|x\| \le 1, \|y\| \le 1} |\langle A^* A x, y \rangle| \\
&= \sup_{x,y \; \|x\| \le 1, \|y\| \le 1} |\langle Ax, Ay \rangle| \\
&\le \|A\|^2,
\end{aligned}
$$

and secondly that

$$\|A\|^2 = \sup_{\|x\| \le 1} \langle Ax, Ax \rangle = \sup_{\|x\| \le 1} \langle A^* A x, x \rangle \le \|A^* A\|.$$

We then observe that

$$
\begin{aligned}
\|P(T)\|^2 &= \|(\bar{P} P)(T)\| \\
&= \sup_{\mu \in \sigma(\bar{P} P)(T)} |\mu| \quad &\text{(using Theorem 6.18)} \\
&= \sup_{\lambda \in \sigma(T)} |(\bar{P} P)(\lambda)| \quad &\text{(using Lemma 8.16)} \\
&= \left(\sup_{\lambda \in \sigma(T)} |P(\lambda)|^2 \right).
\end{aligned}
$$

Step 3. We have defined a map Φ from the set of polynomials into $\mathcal{L}(\mathcal{H})$ defined by

$$P \mapsto \Phi(P) = P(T), \tag{8.4.5}$$

which is continuous:

$$\|\Phi(P)\|_{\mathcal{L}(\mathcal{H})} = \sup_{\lambda \in \sigma(T)} |P(\lambda)|. \tag{8.4.6}$$

The set $\sigma(T)$ is a compact in $\mathbb{R}$ and, using the Stone–Weierstrass theorem (which states the density of the polynomials in $C^0(\sigma(T))$), the map Φ can be uniquely extended to $C^0(\sigma(T))$. We shall denote this extension by Φ also. The properties of Φ are described in the following theorem.

Theorem 8.18 *Let T be a self-adjoint continuous operator on $\mathcal{H}$. Then there exists a unique map Φ from $C^0(\sigma(T))$ into $\mathcal{L}(\mathcal{H})$ with the following properties:*

1.

$$\Phi(f + g) = \Phi(f) + \Phi(g), \quad \Phi(\lambda f) = \lambda \Phi(f);$$
$$\Phi(1) = Id, \qquad\qquad\quad \Phi(\bar{f}) = \Phi(f)^*;$$
$$\Phi(fg) = \Phi(f) \circ \Phi(g).$$

2.

$$\|\Phi(f)\| = \sup_{\lambda \in \sigma(T)} |f(\lambda)|.$$

3. If f is defined by $f(\lambda) = \lambda$, then $\Phi(f) = T$.
4.

$$\sigma(\Phi(f)) = \{f(\lambda) \mid \lambda \in \sigma(T)\}.$$

5. If ψ satisfies $T\psi = \lambda\psi$, then $\Phi(f)\psi = f(\lambda)\psi$.
6. If $f \geq 0$, then $\Phi(f) \geq 0$.

All of these properties can be obtained by showing them first for polynomials P and then extending the properties by continuity to continuous functions. For the last item, note that

$$\Phi(f) = \Phi(\sqrt{f}) \cdot \Phi(\sqrt{f}) = \Phi(\sqrt{f})^* \cdot \Phi(\sqrt{f}).$$

Step 4. We are now ready to introduce measures. Let $\psi \in \mathcal{H}$. Then

$$f \mapsto \langle \psi, f(T)\psi \rangle_{\mathcal{H}} = \langle \psi, \Phi(f)\psi \rangle_{\mathcal{H}}$$

is a nonnegative linear functional on $C^0(\sigma(T))$. By measure theory (Riesz's theorem) (see [Ru1]), there exists a unique measure μ_ψ on $\sigma(T)$, such that

$$\langle \psi, f(T)\psi \rangle_{\mathcal{H}} = \int_{\sigma(T)} f(\lambda) \, d\mu_\psi(\lambda). \tag{8.4.7}$$

This measure is called the spectral measure associated with the vector $\psi \in \mathcal{H}$. This measure is a Borel measure. This means that we can extend the map Φ and (8.4.7) to Borelian functions.

Using the standard Hilbert calculus (that is, the link between sesquilinear forms and quadratic forms), we can also construct, for any x and y in $\mathcal{H}$, a complex measure $d\mu_{x,y}$ such that

$$\langle x, \Phi(f)y \rangle_{\mathcal{H}} = \int_{\sigma(T)} f(\lambda) \, d\mu_{x,y}(\lambda). \tag{8.4.8}$$

Using Riesz's representation theorem (Theorem 3.1), this gives us, when f is bounded, an operator $f(T)$. If $f = 1_{]-\infty,\mu]}$, we recover the operator $E_\mu = f(T)$, which indeed permits us to construct the spectral family announced in Theorem 8.15.   □

Remark 8.19 Modulo some care concerning the domain of the operator, the properties mentioned in the first section of this chapter for operators with a compact resolvent are preserved in the case of an unbounded self-adjoint operator.

8.5 Applications of the spectral theorem

One of the first applications of the spectral theorem (Property 2) is the following property.

Proposition 8.20

$$d(\lambda, \sigma(T)) \, ||x|| \leq ||(T - \lambda)x||, \tag{8.5.1}$$

for all x in $D(T)$.

Remark 8.21 Equation (8.5.1) can be reformulated in the following way:

$$||(T - \lambda)^{-1}|| \leq \frac{1}{d(\lambda, \sigma(T))}. \tag{8.5.2}$$

This proposition is frequently used in the following context. Except for very special cases such as the harmonic oscillator, it is usually difficult to obtain the values of the eigenvalues of an operator explicitly. Consequently, one tries instead to localize these eigenvalues by using approximations. Suppose, for example, that we have found λ_0 and y in $D(T)$ such that

$$||(T - \lambda_0)y|| \leq \epsilon \tag{8.5.3}$$

and $||y|| = 1$. We can then deduce the existence of λ in the spectrum of T such that $|\lambda - \lambda_0| \leq \epsilon$.

Standard examples are the case of Hermitian matrices and the case of the anharmonic oscillator $T := -h^2(d^2/dx^2) + x^2 + x^4$. In the second case, the first eigenfunction of the harmonic oscillator $-h^2(d^2/dx^2) + x^2$ can be used as an approximate eigenfunction y in (8.5.3) with $\lambda_0 = h$. We then find (8.5.3) with $\epsilon = \mathcal{O}(h^2)$.

Another application is, using the property that the spectrum is real, the following inequality:

$$|\operatorname{Im} \lambda|\, ||x|| \leq ||(T - \lambda)x||. \tag{8.5.4}$$

This gives an upper bound on the norm of $(T - \lambda)^{-1}$ in $\mathcal{L}(\mathcal{H})$ of $1/|\operatorname{Im} \lambda|$.

We can also consider the operator $T_\epsilon = -d^2/dx^2 + x^2 + \epsilon x^4$. It can be shown that near each eigenvalue of the harmonic oscillator $(2n+1)$ there exists, when $\epsilon > 0$ is small enough, an eigenvalue $\lambda_n(\epsilon)$ of T_ϵ.

Another good example to analyze is the construction of a sequence of approximate eigenfunctions of the kind considered in Section 1.2. From the construction of u_n such that, with $T = -\Delta$,

$$||(T - \xi^2)u_n||_{L^2(\mathbb{R}^m)} = \mathcal{O}\left(\frac{1}{n}\right),$$

one obtains the result that

$$d(\sigma(T), \xi^2) \leq \frac{C}{n}, \ \forall n \in \mathbb{N}.$$

As $n \to +\infty$, we obtain $\xi^2 \in \sigma(T)$. It is then easy to show that

$$\sigma(P) = [0, +\infty[\,.$$

In fact, it is enough to prove, using the Fourier transform, that for any $b > 0$, $(-\Delta + b)$ has an inverse $(-\Delta + b)^{-1}$ sending L^2 onto H^2.

Here we have followed in a particular case the proof of the following general theorem.

Theorem 8.22 *Let T be a self-adjoint operator. Then $\lambda \in \sigma(T)$ if and only if there exists a sequence $(u_n)_{n \in \mathbb{N}}$, $u_n \in D(T)$, such that $||u_n|| = 1$ and $||(T - \lambda)u_n|| \to 0$ as $n \to +\infty$.*

We leave the proof of the "only if" to the reader, after reading the proof of the next proposition.

Proposition 8.23

$$\sigma(T) = \{\lambda \in \mathbb{R}, \text{ s.t. } \forall \epsilon > 0,\ E(]\lambda - \epsilon, \lambda + \epsilon[) \neq 0\}. \tag{8.5.5}$$

The proof uses, in one direction, the explicit construction of $(T - \lambda)^{-1}$ through Proposition 8.20. If λ and $\epsilon_0 > 0$ are such that $E_{]\lambda-\epsilon_0,\lambda+\epsilon_0[} = 0$, then there exists a continuous function f on $\mathbb{R}$ such that $f(t) = (t - \lambda)^{-1}$ on the support of the measure dE_λ. This permits us to construct the inverse and to show that λ is in the resolvent set of T.

Conversely, let λ be in the set defined by the right-hand side of (8.5.5), later denoted by $\tilde{\sigma}(T)$. For any $n \in \mathbb{N}^*$, choose x_n such that $||x_n|| = 1$ and $E(]\lambda - 1/n, \lambda + 1/n[)x_n = x_n$. Using property 2 of Theorem 8.18, with the function

$t \mapsto f_n(t) = (t - \lambda)1_{]\lambda-1/n,\lambda+1/n[}(t)$, we obtain

$$||(T - \lambda)x_n|| = \left\|(T - \lambda)E\left(\left]\lambda - \frac{1}{n}, \lambda + \frac{1}{n}\right[\right)x_n\right\| \leq \frac{1}{n}||x_n|| = \frac{1}{n}.$$

Applying Proposition 8.20, we obtain

$$\tilde{\sigma}(T) \subset \sigma(T).$$

8.6 Examples of functions of a self-adjoint operator

We encounter various examples of functions f in applications of the spectral theorem.

1. If f is the characteristic function of $]-\infty, \lambda]$, $Y_{]-\infty,\lambda]}$, then $\Phi(f) = f(T)$ becomes $\Phi(f) = E(\lambda)$.
2. If f is the characteristic function of $]-\infty, \lambda[$, $Y_{]-\infty,\lambda[}$, then $f(T)$ is $\Phi(f) = E(\lambda - 0)$.
3. If f is a compactly supported continuous function, then $f(T)$ is an operator whose spectrum is localized in the range of f.
4. If $f_t(\lambda) = \exp(it\lambda)$ with t real, then $f_t(T)$ is a solution of the functional equation

$$(\partial_t - iT)(f(t, T)) = 0,$$
$$f(0, T) = Id.$$

We note here that for all real t, $f_t(T) = \exp(itT)$ is a bounded unitary operator.

5. If $g_t(\lambda) = \exp(-t\lambda)$ with t real and nonnegative, then $g_t(T)$ is a solution of the functional equation

$$(\partial_t + T)(g(t, T)) = 0 \quad \text{for } t \geq 0,$$
$$g(0, T) = Id.$$

We have discussed in Section 8.1 the case of an operator with a compact resolvent. The other case that the beginner needs to understand is of course the case of the free Laplacian $-\Delta$ on $\mathbb{R}^m$. Using the Fourier transform $\mathcal{F}$ (see (2.1.5)), we obtain the operator of multiplication by ξ^2 as the unbounded operator. It is not difficult to define directly the functional calculus, which simply becomes the following for a Borelian function ϕ:

$$\phi(-\Delta) = \mathcal{F}^{-1}\phi(\xi^2)\mathcal{F}. \tag{8.6.1}$$

One possibility is to start from $(-\Delta + 1)^{-1}$, for which this formula is true, and then to follow our previous construction of the functional calculus. Another possibility is to use the formula (8.1.12) and the fact that (8.6.1) is satisfied for $(-\Delta + z)^{-1}$, with $z \in \mathbb{C} \setminus \mathbb{R}$.

The spectral family is then defined by

$$\langle E(\lambda)f, g\rangle_{L^2(\mathbb{R}^m)} = \int_{\xi^2 \leq \lambda} \hat{f}(\xi) \cdot \overline{\hat{g}}(\xi) \, d\xi. \tag{8.6.2}$$

8.7 Spectrum and spectral measures

Another reason for interest in the spectral theorem is that it permits the study of various properties of the spectrum according to the nature of the spectral measure, and this leads to the definition of the continuous spectrum and the pure point spectrum. Let us briefly discuss these notions (without proof).

Starting from a self-adjoint operator T, we define $\mathcal{H}_{pp}$ (a pure point subspace) as the set defined as

$$\mathcal{H}_{pp} = \{\psi \in \mathcal{H} \mid f \mapsto \langle f(T)\psi, \psi\rangle_{\mathcal{H}} \text{ is a pure point measure}\}. \tag{8.7.1}$$

We recall that a measure on X is a pure point measure if

$$\mu(X) = \sum_{x \in X} \mu(\{x\}). \tag{8.7.2}$$

It can be verified that H_{pp} is a closed subspace of $\mathcal{H}$ and that the corresponding orthogonal projection $\Pi_{\mathcal{H}_{pp}}$ satisfies

$$\Pi_{\mathcal{H}_{pp}} D(T) \subset D(T).$$

In this case $T_{/\mathcal{H}_{pp}}$ is naturally defined as an unbounded operator on $\mathcal{H}_{pp}$, and the pure point spectrum of T is defined by

$$\sigma_{pp}(T) = \sigma(T_{/\mathcal{H}_{pp}}). \tag{8.7.3}$$

Similarly, we can define $\mathcal{H}_c$ (a continuous subspace) as the set defined as

$$\mathcal{H}_c = \{\psi \in \mathcal{H} \mid f \mapsto \langle f(T)\psi, \psi \rangle_{\mathcal{H}} \text{ is a continuous measure}\}. \tag{8.7.4}$$

We recall that a measure on X is continuous if

$$\mu(\{x\}) = 0, \ \forall x \in X. \tag{8.7.5}$$

It can be verified that H_c is a closed subspace of $\mathcal{H}$ and that the corresponding orthogonal projection $\Pi_{\mathcal{H}_c}$ satisfies

$$\Pi_{\mathcal{H}_c} D(T) \subset D(T).$$

Moreover, it can be shown that

$$\mathcal{H} = \mathcal{H}_{pp} \oplus \mathcal{H}_c. \tag{8.7.6}$$

In this case $T_{/\mathcal{H}_c}$ is naturally defined as an unbounded operator on $\mathcal{H}_c$, and the continuous spectrum of T is defined by

$$\sigma_c(T) = \sigma(T_{/\mathcal{H}_c}). \tag{8.7.7}$$

Example 8.24 The spectrum of $-\Delta$ is continuous.

We observe that for any $\lambda \geq 0$ and any $f \in L^2(\mathbb{R}^m)$,

$$\lim_{\epsilon \to 0} \langle E_{]\lambda-\epsilon, \lambda+\epsilon[} f, f \rangle = 0.$$

Using (8.6.2), this is a consequence of

$$\lim_{\epsilon \to 0, \, \epsilon > 0} \int_{|\,|\xi|^2 - \lambda| \leq \epsilon} |\hat{f}(\xi)|^2 \, d\xi = 0,$$

which can be verified by use of the dominated convergence theorem.

One can perform an even deeper analysis by using the natural decomposition of the measure given by the Radon–Nikodym theorem. This leads to the notion of the absolutely continuous spectrum and the singularly continuous spectrum.

8.8 Notes

For this rather standard part of spectral theory and for more details, readers are referred to the books by Rudin [Ru1, Ru2], Reed and Simon [RS-I, RS-II, RS-III, RS-IV], Huet [Hu], and Lévy-Bruhl [Le-Br]. In particular, Theorem 8.12 can be found in Reed and Simon, Vol. 1 [RS-I], and Theorem 8.14 can be found in [Ru2]. For Theorem 8.15, readers are referred to [Hu], [Le-Br], and [DaLi] for detailed proofs and to [RS-I] for another proof, which inspired the presentation given here. Another interesting proof is based on (8.1.12) and presented in Davies's book [Dav1]. This formula is due to Dynkin but was popularized by Helffer and Sjöstrand in the context of spectral theory, leading many authors to call it the Helffer–Sjöstrand formula.

For Section 8.7, readers are referred to [RS-I]. The Radon–Nikodym theorem can be found in [Ru2].

8.9 Exercises

Exercise 8.1 Show that the spectrum of the self-adjoint operator in $L^2(\mathbb{R}^2)$ $P = D_x^2 + x^2 + D_y^2$ (with domain suitably defined!) is $[1, +\infty[$. With (8.6.2) in mind, describe the spectral family E_λ associated with P. Does P admit eigenfunctions?

Exercise 8.2 (Airy operator.) Show that the spectrum of $D_x^2 + x$ (with domain suitably defined) on $L^2(\mathbb{R})$ is $]-\infty, +\infty[$.

Hint: Use the Fourier transform.

Exercise 8.3 Let $G = (V, E)$ be a graph, where V is the collection of vertices and $E \subset V \times V$ is the set of edges (that connect pairs of vertices), with $(u, v) \in E \Leftrightarrow (v, u) \in E$. We define a graph Laplacian (or discrete Laplacian) as the operator on $\ell^2(V)$ defined by

$$f \mapsto (\Delta^G f)(v) = \sum_{u:(u,v)\in E} f(u), \quad v \in V,$$

where

$$\ell^2(V) = \left\{ f : V \to \mathbb{C} : \sum_{v\in V} |f(v)|^2 < +\infty \right\}, \quad \langle f, g \rangle := \sum_{v\in V} f(v)\overline{g(v)}.$$

(a) Compute the spectrum of the graph Laplacian Δ^G in the case of a triangle.

(b) Compute the spectrum of the graph Laplacian Δ^G for the graph G naturally associated with a square lattice $V := \mathbb{Z}^2$,

$$f \mapsto (\Delta^G f)(v) = \sum_{|u-v|=1} f(u), \quad v \in V.$$

(c) Does Δ^G have eigenvalues?

Hint: Use the unitary transform

$$U : \ell^2(\mathbb{Z}^2) \to L^2([0, 2\pi] \times [0, 2\pi]), \ U(f_{n_1,n_2})(\theta_1, \theta_2)$$

$$= \frac{1}{2\pi} \sum_{(n_1,n_2)\in\mathbb{Z}^2} f_{n_1,n_2} e^{i(n_1\theta_1 + n_2\theta_2)}.$$

9

Essentially self-adjoint operators

In most of the examples that we have presented as applications of the abstract theory, the operators considered were associated with differential operators. These differential operators are naturally defined on $C_0^\infty(\Omega)$ or $\mathcal{D}'(\Omega)$. Most of the time (for suitable potentials increasing slowly at ∞), they are also defined (when $\Omega = \mathbb{R}^m$) on $\mathcal{S}(\mathbb{R}^m)$ or $\mathcal{S}'(\mathbb{R}^m)$. It is important to understand how the abstract point of view can be related to the point of view of partial differential equations and to find out whether there is a natural way to introduce a self-adjoint problem. The theory of essential self-adjointness gives a clear understanding of the problem and permits us to propose assumptions under which the associated self-adjoint operator is essentially unique.

9.1 Symmetric operators and self-adjoint extensions

More mathematically, the question is to decide whether, starting from a symmetric operator T whose domain $D(T) = \mathcal{H}_0$ is dense in $\mathcal{H}$, there exists a unique self-adjoint extension T^{ext} of T. We recall that this means that $D(T) \subset D(T^{\text{ext}})$ and $T^{\text{ext}}u = Tu$, $\forall u \in D(T)$. This leads to the following definition.

Definition 9.1 A symmetric operator T with domain $\mathcal{H}_0$ is called essentially self-adjoint if its closure is self-adjoint.

It can easily be verified (see Proposition 2.9 in conjunction with the definition of $\overline{T}$) that $T^* = \overline{T}^*$, and we recall that $T^{**} = \overline{T}$.

We now observe the following.

Proposition 9.2 *Any closed symmetric extension of T is a restriction of T^*.*

Proof Let S be a closed symmetric extension of T.

We have in fact $T \subset S \subset S^*$. Returning to the definition of the adjoint, it can be seen immediately that if $T \subset S$, then $S^* \subset T^*$. Hence we obtain $S \subset T^*$. □

In particular, we can state the following proposition.

Proposition 9.3 *If T is essentially self-adjoint, then its self-adjoint extension is unique.*

Suppose that S is a self-adjoint extension of T. Then S is closed and, being an extension of T, it is also an extension of its smallest extension $\overline{T}$. Thus,

$$\overline{T} \subset S = S^* \subset (\overline{T})^* = \overline{T},$$

and so $S = \overline{T}$.

Examples 9.4 Here, we give a list of examples and counterexamples that will be analyzed later.

1. The differential operator $-\Delta$ with domain $\mathcal{H}_0 = C_0^\infty(\mathbb{R}^m)$ is essentially self-adjoint (see later).
2. The differential operator $-\Delta + |x|^2$ with domain $\mathcal{H}_0 = C_0^\infty(\mathbb{R}^m)$ or $\mathcal{H}_0 = \mathcal{S}(\mathbb{R}^m)$ is essentially self-adjoint and, consequently, admits a unique self-adjoint extension (the harmonic oscillator). This extension is the same for the two operators. The domain can be described explicitly as

$$B^2(\mathbb{R}^m) = \{u \in L^2(\mathbb{R}^m) \mid x^\alpha D_x^\beta u \in L^2(\mathbb{R}^m), \forall \alpha, \beta \in \mathbb{N}^m \text{ with } |\alpha| + |\beta| \leq 2\}.$$

3. The differential operator $-\Delta$ with domain $C_0^\infty(\Omega)$ (where Ω is an open bounded set with a smooth boundary) is not essentially self-adjoint. There exist many self-adjoint extensions related to the choice of boundary problems. As we have seen before, we have already met two such extensions:
   - the Dirichlet realization, whose domain is the set $H_0^1(\Omega) \cap H^2(\Omega)$;
   - the Neumann realization, whose domain is the set

$$\{u \in H^2(\Omega) \mid (\partial u / \partial \nu)_{/\partial\Omega} = 0\}.$$

4. The Laplace–Beltrami operator on a compact manifold M with domain $C^\infty(M)$ is essentially self-adjoint on $L^2(M)$. The domain of the self-adjoint extension is $H^2(M)$ (which can be described using local charts as the space of the distributions that are locally in H^2).

9.2 Basic criteria

We now give some criteria for verifying that an operator is essentially self-adjoint. As already mentioned, one can prove essential self-adjointness by proving that the minimal closed extension $T_{\min} := \overline{T}$ coincides with T^*.

We can characterize self-adjointness using the following general criterion.

Theorem 9.5 *Let T be a closed symmetric operator. Then the following statements are equivalent:*

1. *T is self-adjoint.*
2. *$N(T^* \pm i) = \{0\}$.*
3. *$R(T \pm i) = \mathcal{H}$.*

Proof

1 implies 2

This property has already been observed (because $T = T^*$ and $\pm i \notin \sigma(T)$).

2 implies 3

We first observe that the property that $N(T^* + i) = \{0\}$ implies that $R(T - i)$ is dense in $\mathcal{H}$. Note that the converse is also true. To obtain 3, it remains to show that $R(T - i)$ is closed. But, for all ϕ in $D(T)$, we have (using the symmetry of T)

$$||(T - i)\phi||^2 = ||T\phi||^2 + ||\phi||^2. \tag{9.2.1}$$

If ϕ_n is a sequence in $D(T)$ such that $(T + i)\phi_n$ converges to some ψ_∞, then the previous identity shows that ϕ_n is a Cauchy sequence, so there exists ϕ_∞ such that $\phi_n \to \phi_\infty$ in $\mathcal{H}$. But $T\phi_n = (T + i)\phi_n - i\phi_n$ is convergent and, using the fact that the graph is closed, we obtain the result that $\phi_\infty \in D(T)$ and $T\phi_\infty = \psi_\infty - i\phi_\infty$.

3 implies 1

Let $\phi \in D(T^*)$. Let $\eta \in D(T)$ such that $(T - i)\eta = (T^* - i)\phi$. Since T is symmetric, we have also $(T^* - i)(\eta - \phi) = 0$. But, if $(T + i)$ is surjective, then $(T^* - i)$ is injective and we obtain $\phi = \eta$. This proves that $\phi \in D(T)$. $\square$

Here we have used and proved during the proof of the assertion "2 implies 3" the following lemma.

Lemma 9.6 *If T is closed and symmetric, then $R(T \pm i)$ is closed.*

Theorem 9.5 leads as a corollary to the following criterion for essential self-adjointness.

Corollary 9.7 *Let A, with domain $D(A)$, be a symmetric operator. Then the following statements are equivalent:*

1. *A is essentially self-adjoint.*
2. $N(A^* + i) = \{0\}$ *and* $N(A^* - i) = \{0\}$.
3. $R(A + i)$ *and* $R(A - i)$ *are dense in* $\mathcal{H}$.

Essentially, we have to apply Theorem 9.5 to $\overline{A}$, observe in addition that $\overline{A}$ is symmetric, and use Lemma 9.6.

In the same spirit, we have the following theorem in the semibounded case.

Theorem 9.8 *Let T be a positive, symmetric operator. Then the following statements are equivalent:*

1. *T is essentially self-adjoint.*
2. $N(T^* + b) = \{0\}$ *for some* $b > 0$.
3. $R(T + b)$ *is dense for some* $b > 0$.

The proof is essentially the same as that for Corollary 9.7, where we observe that if T is nonnegative, then the following trivial estimate is a good substitute for (9.2.1):

$$\langle (T + b)u, u \rangle \geq b||u||^2. \tag{9.2.2}$$

Note that the proof also gives the property that $N(T^* + b) = \{0\}$ for any $b > 0$.

Example 9.9 (The free Laplacian.) The operator $-\Delta$ with domain C_0^∞ is essentially self-adjoint. Its self-adjoint extension is $-\Delta$ with domain H^2.

9.3 The Kato–Rellich theorem

We shall consider the case where $P = -\Delta + V$ when V is regular and tends to 0 as $|x| \to +\infty$. Here V can be considered as a perturbation of the Laplacian. We can then apply a general theorem due to Kato and Rellich.

Theorem 9.10 *Let A be a self-adjoint operator, and let B be a symmetric operator whose domain contains $D(A)$. Assume the existence of a and b such that $0 \leq a < 1$ and $b \geq 0$ such that*

$$||Bu|| \leq a\,||Au|| + b\,||u||, \tag{9.3.1}$$

for all $u \in D(A)$. Then $A + B$ is self-adjoint on $D(A)$. If A is essentially self-adjoint on[1] *$D \subset D(A)$, then $A + B$ has the same property.*

---

[1] By this, we mean that the closure of $A_{/D}$ is A.

Proof

Step 1. We start from the following identity, which uses only the fact that $(A + B)$ with domain $D(A)$ is symmetric:

$$||(A + B \pm i\lambda)u||^2 = ||(A + B)u||^2 + \lambda^2 ||u||^2, \quad \forall u \in D(A). \quad (9.3.2)$$

By the triangle inequality and the symmetry of $A + B$, we obtain, for any real $\lambda > 0$ and any $u \in D(A)$,

$$\begin{aligned}
\sqrt{2} ||(A + B - i\lambda)u|| &\geq ||(A + B)u|| + \lambda ||u|| \\
&\geq ||Au|| - ||Bu|| + \lambda ||u|| \quad (9.3.3) \\
&\geq (1 - a) ||Au|| + (\lambda - b) ||u||.
\end{aligned}$$

We now choose $\lambda > b$.

Step 2. In this step, we show that $(A + B)$ with domain $D(A)$ is closed. If we start from a pair (u_n, f_n) with $u_n \in D(A)$ and $f_n = (A + B)u_n$ such that $(u_n, f_n) \to (u, f)$ in $\mathcal{H}$, then we find from (9.3.3) that Au_n is a Cauchy sequence in $\mathcal{H}$. Since A is closed, we obtain $u \in D(A)$ and the existence of g such that $Au_n \to g = Au$.

Now, from (9.3.2) and (9.3.1), we obtain also the result that Bu_n is a Cauchy sequence and there exists v such that $Bu_n \to v$ in $\mathcal{H}$.

We claim that $Bu = v$. We have in fact, for any $h \in D(A)$,

$$\langle v, h \rangle_{\mathcal{H}} = \lim_{n \to +\infty} \langle Bu_n, h \rangle_{\mathcal{H}} = \lim_{n \to +\infty} \langle u_n, Bh \rangle_{\mathcal{H}} = \langle u, Bh \rangle_{\mathcal{H}} = \langle Bu, h \rangle_{\mathcal{H}}.$$

Using the density of $D(A)$, we obtain $v = Bu$. (We could also have used the fact that B is closable).

We conclude by observing that $(A+B)u = f$ (with $f = g+v$), as expected.

Step 3. In order to apply Theorem 9.5, we have to show that $(A + B \pm i\lambda)$ is surjective. The main element in the proof is the following lemma.

Lemma 9.11 *For $\lambda > 0$ large enough, we have*

$$||B(A \pm i\lambda)^{-1}|| < 1. \quad (9.3.4)$$

Proof We observe that for $u \in D(A)$,

$$||(A \pm i\lambda)u||^2 = ||Au||^2 + \lambda^2 ||u||^2. \quad (9.3.5)$$

For $u \in D(A)$, we have, using (9.3.5) twice and then (9.3.1),

$$\|Bu\| \leq a\,\|Au\| + b\,\|u\|$$

$$\leq a\,\|(A+i\lambda)u\| + \frac{b}{\lambda}\,\|(A+i\lambda)u\| \qquad (9.3.6)$$

$$\leq \left(a + \frac{b}{\lambda}\right)\,\|(A+i\lambda)u\|.$$

It is then enough to choose $\lambda > 0$ large enough such that

$$\left(a + \frac{b}{\lambda}\right) < 1.$$

Writing

$$A + B - i\lambda = [I + B(A - i\lambda)^{-1}](A - i\lambda), \qquad (9.3.7)$$

it is easy to deduce the surjectivity of $A + B - i\lambda$ using the lemma and the surjectivity of $(A - i\lambda)$. $\qquad\square$

Application   As an application, let us treat the case of the Schrödinger operator with a Coulomb potential.

Proposition 9.12 *The operator* $-\Delta - 1/|x|$ *with domain* $C_0^\infty(\mathbb{R}^3)$ *is essentially self-adjoint.*

We recall that the operator is well defined because $1/r$ belongs to $L^2_{\mathrm{loc}}(\mathbb{R}^3)$. First, we observe a Sobolev-type inequality, as follows.

Lemma 9.13 *There exists a constant C such that for all* $u \in H^2(\mathbb{R}^3)$, *all* $a > 0$, *and all* $x \in \mathbb{R}^3$, *we have*

$$|u(x)| \leq C(a\,\|\Delta u\|_0 + a^{-3}\,\|u\|_0).$$

We can start by proving the inequality[2] for all u with $x = 0$ and $a = 1$, and then use translation and dilation.

Secondly, we show that the potential $V = -1/r$ is a perturbation of the Laplacian. There exists in fact a constant C such that for all $u \in H^2(\mathbb{R}^3)$ and all $b > 0$, we have

$$\|Vu\|_0 \leq C(b\,\|\Delta u\|_0 + b^{-3}\,\|u\|_0).$$

---

[2] In this case, this is just the Sobolev injection theorem for $H^2(\mathbb{R}^3)$ into $C_b^0(\mathbb{R}^3)$, where $C_b^0(\mathbb{R}^3)$ is the space of continuous bounded functions.

For the proof of this, we observe that for any $R > 0$,

$$\int V(x)^2 |u(x)|^2 \, dx = \int_{|x| \leq R} V(x)^2 |u(x)|^2 \, dx + \int_{|x| \geq R} V(x)^2 |u(x)|^2 \, dx.$$

We treat the first term on the right-hand side using a Sobolev-type inequality,

$$\int_{|x| \leq R} V(x)^2 |u(x)|^2 \, dx \leq \left(\sup_{|x| \leq R} |u(x)|^2 \, dx \right) \int_{|x| \leq R} V(x)^2 \, dx,$$

and treat the second term using the trivial estimate

$$\int_{|x| \geq R} V(x)^2 |u(x)|^2 \, dx \leq \left(\sup_{|x| \geq R} |V(x)| \right)^2 \int_{|x| \geq R} |u(x)|^2 \, dx.$$

Using the Sobolev inequality (actually it is enough to take $a = 1$), we finally obtain

$$\|Vu\|^2 \leq C \, R \, (\|\Delta u\|^2 + \|u\|^2) + \frac{1}{R^2} \|u\|^2. \tag{9.3.8}$$

We obtain the expected estimate by considering R small enough.

Remark 9.14 We note that the same proof also shows that $-\Delta + V$ is essentially self-adjoint, starting from $C_0^\infty(\mathbb{R}^3)$, if $V = V_1 + V_2$ with $V_1 \in L^2(\mathbb{R}^3)$ and $V_2 \in L^\infty(\mathbb{R}^3)$.

9.4 Other criteria for self-adjointness of Schrödinger operators

In this section, we present some criteria that are specific to the Schrödinger case.

The first theorem is adapted to operators that are already known to be positive on $C_0^\infty(\mathbb{R}^m)$.

Theorem 9.15 *A Schrödinger operator $T = -\Delta + V$ on $\mathbb{R}^m$ associated with a C^0 potential V that is semibounded on $C_0^\infty(\mathbb{R}^m)$ is essentially self-adjoint. In other words, the Friedrichs extension is the unique self-adjoint extension starting from $C_0^\infty(\mathbb{R}^m)$.*

This theorem is complementary to the second theorem (Theorem 9.18) that is stated in this section because we do not have to assume the nonnegativity of the potential but only the semiboundedness of the operator T.

Proof Let T be our operator. Possibly by adding a constant, we can assume that

$$\langle Tu, u \rangle_{\mathcal{H}} \geq ||u||^2, \ \forall u \in C_0^\infty(\mathbb{R}^m). \tag{9.4.1}$$

Of course, this inequality can be rewritten in the form

$$||\nabla u||^2 + \int_{\mathbb{R}^m} V(x) |u(x)|^2 \, dx \geq ||u||^2, \ \forall u \in C_0^\infty(\mathbb{R}^m).$$

In this form, the inequality can be extended to the elements of $H_{\text{comp}}^1(\mathbb{R}^m)$, corresponding to the distributions of $H^1(\mathbb{R}^m)$ with compact support:

$$||\nabla u||^2 + \int_{\mathbb{R}^m} V(x)|u(x)|^2 \, dx \geq ||u||^2, \ \forall u \in H_{\text{comp}}^1(\mathbb{R}^m). \tag{9.4.2}$$

According to the general criterion for essential self-adjointness (see Theorem 9.8), it is enough to verify that $R(T)$ is dense. Let us show this property.

Let $f \in L^2(\mathbb{R}^m)$, such that

$$\langle f, Tu \rangle_{\mathcal{H}} = 0, \ \forall u \in C_0^\infty(\mathbb{R}^m). \tag{9.4.3}$$

We have to show that $f = 0$. Because T is real, we can assume that f is real. We first observe that (9.4.3) implies that $(-\Delta + V)f = 0$ in $\mathcal{D}'(\mathbb{R}^m)$. A standard regularity theorem for the Laplacian implies that $f \in H_{\text{loc}}^2(\mathbb{R}^m)$. More precisely, this is the property that $f \in L_{\text{loc}}^2(\mathbb{R}^m)$, $\Delta f \in H_{\text{loc}}^{-1}(\mathbb{R}^m)$ implies that $f \in H_{\text{loc}}^1(\mathbb{R}^m)$, together with the property that $f \in L_{\text{loc}}^2(\mathbb{R}^m)$, $\Delta f \in L_{\text{loc}}^2(\mathbb{R}^m)$ implies that $f \in H_{\text{loc}}^2(\mathbb{R}^m)$.

We now introduce a family of cutoff functions ζ_k defined by

$$\zeta_k := \zeta(x/k), \ \forall k \in \mathbb{N}, \tag{9.4.4}$$

where ζ is a C^∞ function satisfying $0 \leq \zeta \leq 1$, $\zeta = 1$ on $B(0, 1)$, and supp $\zeta \subset B(0, 2)$. For any $u \in C^\infty$ and any $f \in H_{\text{loc}}^2$, we have the identity

$$\int \nabla(\zeta_k f) \cdot \nabla(\zeta_k u) \, dx + \int \zeta_k(x)^2 V(x) u(x) f(x) \, dx$$

$$= \int |(\nabla \zeta_k)(x)|^2 \, u(x) f(x) \, dx$$

$$+ \sum_{i=1}^m \int \left(f(\partial_{x_i} u) - u(\partial_{x_i} f) \right)(x) \zeta_k(x)(\partial_{x_i} \zeta_k)(x) \, dx$$

$$+ \langle f(x), T\zeta_k^2 u \rangle. \tag{9.4.5}$$

When f satisfies (9.4.3), we obtain

$$\int_{\mathbb{R}^m} \nabla(\zeta_k f) \cdot \nabla(\zeta_k u)\, dx + \int \zeta_k(x)^2 V(x) u(x) f(x)\, dx$$
$$= \int |(\nabla \zeta_k)(x)|^2\, u(x) f(x)\, dx$$
$$+ \sum_{i=1}^{m} \int \left(f(\partial_{x_i} u) - u(\partial_{x_i} f) \right)(x) \zeta_k(x)(\partial_{x_i} \zeta_k)(x)\, dx, \quad (9.4.6)$$

for all $u \in C^\infty(\mathbb{R}^m)$.

This formula can be extended to functions $u \in H^1_{\text{loc}}$. In particular, we can take $u = f$. We obtain

$$\langle \nabla(\zeta_k f),\, \nabla(\zeta_k f) \rangle + \int \zeta_k^2 V(x)\, |f(x)|^2\, dx = \int |\nabla \zeta_k|^2\, |f(x)|^2 dx.$$
$$(9.4.7)$$

Using (9.4.1) and (9.4.7) and taking the limit $k \to +\infty$, we obtain

$$\|f\|^2 = \lim_{k \to +\infty} \|\zeta_k f\|^2$$
$$\leq \lim_{k \to +\infty} \sup \left(\langle \nabla(\zeta_k f),\, \nabla(\zeta_k f) \rangle + \int \zeta_k^2 V(x)\, |f(x)|^2\, dx \right)$$
$$= \lim_{k \to +\infty} \sup \int f(x)^2\, |(\nabla \zeta_k)(x)|^2\, dx = 0. \quad (9.4.8)$$

This proves the theorem. $\qquad \square$

Remark 9.16 When V is C^∞, we obtain in the above proof the result that $f \in C^\infty$, and we can immediately prove (9.4.7) without going through the previous discussion.

Examples 9.17

1. If $V \geq 0$ and C^∞, then the Schrödinger operator $-\Delta + V$ with domain $C_0^\infty(\mathbb{R}^m)$ is essentially self-adjoint. The operator $-\Delta + V$ is in fact positive.
2. If ϕ is C^∞ on $\mathbb{R}^m$, then the operators $-\Delta + |\nabla \phi|^2 \pm \Delta \phi$ are essentially self-adjoint. They are in fact nonnegative on $C_0^\infty(\mathbb{R}^m)$. They can be written in the form $\sum_j Z_j^* Z_j$, with $Z_j = \partial_{x_j} \mp \partial_{x_j} \phi$. These operators appear naturally in statistical mechanics.

Let us now mention, without proof, a quite general theorem due to Kato.

Theorem 9.18 *Let V in $L^2_{\text{loc}}(\mathbb{R}^m)$ such that $V \geq 0$ almost everywhere on $\mathbb{R}^m$. Then $-\Delta + V$ with domain $C^\infty_0(\mathbb{R}^m)$ is essentially self-adjoint.*

This theorem is based on the so-called Kato's inequality (see below).

Remark 9.19 The above theorem may be extended to the case of the Schrödinger operator with a regular magnetic potential A (see Section 7.4).

A useful tool for obtaining stronger results about essential self-adjointness is Kato's inequality. We present it in the magnetic version in Theorem 9.21. Let us start, however, with the case without a magnetic field.

Theorem 9.20 *(Kato's inequality.) Let $f \in L^1_{\text{loc}}(\mathbb{R}^m)$ such that $\Delta f \in L^1_{\text{loc}}(\mathbb{R}^m)$. Then we have the inequality*

$$\Delta |f| \geq \text{Re}\{\text{sign}(f)\,\Delta f\}, \tag{9.4.9}$$

almost everywhere, where sign f is defined in (7.2.3).

The proof of Theorem 9.20 follows the same steps as the proof of Proposition 7.7. That is, we first consider smooth functions f and the regularized absolute value

$$|z|_\epsilon = \sqrt{|z|^2 + \epsilon^2} - \epsilon,$$

and calculate directly. We then consider a sequence f_δ of smooth approximations to f. Taking first δ and then ϵ to zero, we obtain the desired inequality. We leave the details to the reader.

Theorem 9.21 *(Kato's magnetic inequality.) Let $\mathbf{A} \in C^1(\mathbb{R}^m, \mathbb{R}^m)$. Then, for all $f \in L^1_{\text{loc}}(\mathbb{R}^m)$ with $(-i\nabla + \mathbf{A})^2 f \in L^2_{\text{loc}}(\mathbb{R}^m)$, we have the inequality*

$$\Delta |f| \geq -\text{Re}\{\text{sign}(f)\,(-i\nabla + \mathbf{A})^2 f\}, \tag{9.4.10}$$

where sign f is defined in (7.2.3).

Proof We give only a proof under the extra regularity assumption that $\mathbf{A} \in C^2(\mathbb{R}^m)$. In this case the assumption $(-i\nabla + \mathbf{A})^2 f \in L^2_{\text{loc}}(\mathbb{R}^m)$ and the standard regularity property of second-order elliptic operators imply that $f \in H^2_{\text{loc}}(\mathbb{R}^m)$. In particular, we have

$$\Delta f, \nabla f \in L^1_{\text{loc}}(\mathbb{R}^m). \tag{9.4.11}$$

Suppose now that u is smooth. Then we can calculate as follows:

$$\nabla |u|_\epsilon = \frac{\text{Re}\,\{\overline{u}\,\nabla u\}}{\sqrt{|u|^2 + \epsilon^2}} = \frac{\text{Re}\,\{\overline{u}\,(\nabla + i\mathbf{A})u\}}{\sqrt{|u|^2 + \epsilon^2}}. \tag{9.4.12}$$

We therefore find

$$\begin{aligned}
\sqrt{|u|^2 + \epsilon^2}\,\Delta |u|_\epsilon &= \text{div}\,(\sqrt{|u|^2 + \epsilon^2}\,\nabla |u|_\epsilon) - |\nabla |u|_\epsilon|^2 \\
&= \text{Re}\,\left\{\overline{\nabla u} \cdot (\nabla + i\mathbf{A})u + \overline{u}\,\text{div}\,((\nabla + i\mathbf{A})u)\right\} - |\nabla |u|_\epsilon|^2 \\
&= |(\nabla + i\mathbf{A})u|^2 - |\nabla |u|_\epsilon|^2 \\
&\quad + \text{Re}\,\left\{i\mathbf{A}\overline{u} \cdot (\nabla + i\mathbf{A})u + \overline{u}\,\text{div}\,((\nabla + i\mathbf{A})u)\right\}.
\end{aligned} \tag{9.4.13}$$

By (9.4.12) and using the Cauchy–Schwarz inequality, we obtain

$$|(\nabla + i\mathbf{A})u|^2 \geq |\nabla |u|_\epsilon|^2.$$

So (9.4.13) implies that, for a smooth u,

$$\Delta |u|_\epsilon \geq \text{Re}\,\frac{\overline{u}\,(\nabla + i\mathbf{A})^2 u}{\sqrt{|u|^2 + \epsilon^2}}. \tag{9.4.14}$$

The end of the proof now follows the same lines as the proof of Proposition 7.7, i.e., (9.4.14) holds for suitably smoothed versions f_δ of f. By taking the limit $\delta \to 0$ followed by the limit $\epsilon \to 0$, in the sense of distributions, we arrive at (9.4.10).

If we know only that $\mathbf{A} \in C^1$, we cannot immediately conclude (9.4.11); a proof in this case will not be given. $\qquad\square$

By the above criterion, it is enough to show that if $u \in L^2$ satisfies $(-\Delta + V + 1)u = 0$, then $u = 0$. We first see that

$$(-\Delta + 1)\,|u| \leq 0. \tag{9.4.15}$$

We have in fact $(\Delta - 1)\,u = Vu$ and, using Kato's inequality,

$$\Delta |u| \geq \text{Re}\,\{\text{sign}\,(u)\,\Delta u\} = (V + 1)\,|u|.$$

We then use the fact that $(-\Delta + 1)^{-1}$ is positivity-improving. Although the result is general, we shall first give a proof when $m = 3$. The operator is associated with the distribution kernel

$$K(x, y) = \frac{1}{4\pi |x - y|} e^{-|x-y|}. \tag{9.4.16}$$

We then obtain a contradiction unless $|u| = 0$.

More generally, we have $K(x, y) = g(x - y)$, with

$$g(x) = (2\pi)^{-m} \int_{\mathbb{R}^m} e^{ix\cdot\xi} (|\xi|^2 + 1)^{-1} d\xi.$$

We can then observe that g is radial:

$$g(x) = h(|x|),$$

with, for $t \geq 0$,

$$h(t) = (2\pi)^{-m} \int_{\mathbb{R}\times\mathbb{R}^{m-1}} e^{it\sigma} (\sigma^2 + |\xi'|^2 + 1)^{-1} d\sigma \, d\xi'.$$

Hence

$$h(t) = c_m \int_{-\infty}^{+\infty} \int_0^{+\infty} e^{it\sigma} (\sigma^2 + r^2 + 1)^{-1} r^{m-2} \, d\sigma \, dr$$

$$= \tilde{c}_m \int_0^{+\infty} e^{-t(r^2+1)^{1/2}} (r^2 + 1)^{-1/2} r^{m-2} \, dr. \tag{9.4.17}$$

Hence $h(t) > 0$, and this proves the positivity-preserving property. When $m = 3$, we recover (9.4.16).

9.5 Notes

In Section 9.1, although it is not needed in this book, note that the converse of Proposition 9.3 is true. See, for example, the book [Ro], in which essential self-adjointness is defined differently. In Section 9.4, the first theorem seems to be due to Rellich (see [Sima]), and we have presented it in the easy case where the potential is regular. The second theorem permits us to treat singular potentials and is due to Kato (see [HiSi] or [Ro]).

For Theorem 9.18, readers are referred to [Ro], for example. The details of the proof of (9.4.11) in the magnetic case can be found in [RS-II, Vol. 2, section X.4].

9.6 Exercises

Exercise 9.1 Let $A(x_1, x_2) = (A_1(x_1, x_2), A_2(x_1, x_2))$ be a C^∞ vector field on $\mathbb{R}^2$. Let V be a C^∞ positive function on $\mathbb{R}^2$. Let

$$P := (D_{x_1} - A_1(x_1, x_2))^2 + (D_{x_2} - A_2(x_1, x_2))^2 + V(x_1, x_2)$$

be the differential operator defined on $C_0^\infty(\mathbb{R}^2)$. We recall that $D_{x_j} = (1/i)\partial_{x_j}$.

(a) Show that P admits a self-adjoint extension in $L^2(\mathbb{R}^2)$.
(b) Show that P is essentially self-adjoint.

Exercise 9.2 Consider on $C_0^\infty(\mathbb{R}^2)$ the operator

$$P_0 := (D_{x_1} - x_2)^2 + (D_{x_2} - x_1)^2.$$

Show that its natural self-adjoint extension P is unitarily equivalent to the operator $-\Delta$ (of domain H^2). Determine its spectrum and its essential spectrum.

Exercise 9.3 Returning to Exercise 4.1, answer the following additional questions:

(a) Show that $\mathcal{D}_0$ admits a self-adjoint extension $\mathcal{D}_1$ in $L^2(\mathbb{R}^2; \mathbb{C}^2)$, and determine its domain.
(b) Determine the spectrum of $\mathcal{D}_1$.
(c) Suppose that for all $x \in \mathbb{R}^2$, $V(x)$ is a 2×2 Hermitian matrix, with bounded C^∞ coefficients. Show that $\mathcal{D}_V = \mathcal{D}_0 + V$ admits a self-adjoint extension and determine its domain.

Exercise 9.4 Discuss, as a function of $\alpha \geq 0$, the possibility of associating a self-adjoint operator on $L^2(\mathbb{R}^3)$ with the differential operator defined on $C_0^\infty(\mathbb{R}^3)$ by $-\Delta - r^{-\alpha}$, with $r = \sqrt{x^2 + y^2 + z^2}$.

10

The discrete spectrum and essential spectrum

As many examples and exercises have shown, the spectrum of a self-adjoint operator (and, a fortiori, of a non-self-adjoint operator) can take various forms. We have already mentioned how we can use the existence of the spectral measure to decompose a spectrum into a union of different spectra. Here we shall focus on the description of another possible decomposition into two spectra, bearing in mind that one of these spectra (the essential spectrum) is more stable with respect to suitable perturbations and that the other (the discrete spectrum) corresponds more to what we learned in our first years of university about Hermitian matrices.

10.1 Discrete spectrum

We gave a characterization of the spectrum in Proposition 8.23. We now complete this characterization by introducing some other spectra.

Definition 10.1 If T is a self-adjoint operator, we call the discrete spectrum of T the set

$$\sigma_{\text{disc}}(T) = \{\lambda \in \sigma(T) \text{ s.t. } \exists \epsilon > 0, \dim R(E(]\lambda - \epsilon, \lambda + \epsilon[)) < +\infty\}.$$

With this new definition, we can say that for a self-adjoint operator with a compact resolvent, the spectrum is reduced to the discrete spectrum. For a compact self-adjoint operator, the spectrum is discrete outside 0. We see in this case that the discrete spectrum is not closed.

Equivalently, we can now give another characterization, as follows.

Proposition 10.2 *Let T be a self-adjoint operator. A real λ is in the discrete spectrum if and only if λ is an isolated point in $\sigma(T)$ and an eigenvalue of finite multiplicity (that is, with a corresponding eigenspace of finite dimension).*

Proof This is based on two points.

(a) If $\lambda \in \sigma_{\text{disc}}(T)$, we can immediately see that there exists ϵ_0 such that, $\forall \epsilon \in]0, \epsilon_0[$, $E_{]\lambda-\epsilon,\lambda+\epsilon[}$ becomes a projector independent of ϵ with finite range. This is actually the projector $\Pi_\lambda = 1_{\{\lambda\}}(T)$, and we observe moreover that $E_{]\lambda,\lambda+\epsilon_0[} = 0$ and $E_{]\lambda-\epsilon_0,\lambda[} = 0$. This shows that λ is an isolated point in $\sigma(T)$.

(b) If λ is an isolated point, we immediately obtain, using the spectral representation of T, the result that if $x = \Pi_\lambda x$ ($x \neq 0$), then x is an eigenfunction of T. Moreover, we easily obtain the result that $(T - \lambda)$ is invertible on $R(I - \Pi_\lambda)$. One can in fact find a continuous bounded f such that $f(T)(T - \lambda)(I - \Pi_\lambda) = (I - \Pi_\lambda)$. It can also be observed that in this case the range of Π_λ is an eigenspace. $\qquad\square$

Remark 10.3 The discrete spectrum is not a closed set! If we consider, in $\mathbb{R}^3$, the Schrödinger operator with a Coulomb potential, the discrete spectrum is a sequence of eigenvalues tending to 0, but 0 does not belong to the discrete spectrum.

10.2 Essential spectrum

Definition 10.4 The essential spectrum is the complement in the spectrum of the discrete spectrum.

Intuitively, a point in the essential spectrum corresponds to either:

- a point in the continuous spectrum;
- a limit point of a sequence of eigenvalues with finite multiplicity; or
- an eigenvalue of infinite multiplicity.

Since the discrete spectrum is composed of isolated points, we obtain the following proposition.

Proposition 10.5 *The essential spectrum of a self-adjoint operator T is closed in $\mathbb{R}$.*

10.3 Basic examples

1. The essential spectrum of a compact self-adjoint operator is reduced to 0.
2. The essential spectrum of an operator with a compact resolvent is empty.
3. The Laplacian on $\mathbb{R}^m$ $-\Delta$ is a self-adjoint operator on $L^2(\mathbb{R}^m)$ whose domain is the Sobolev space $H^2(\mathbb{R}^m)$. The spectrum is continuous and

equal to $\overline{\mathbb{R}^+}$. The essential spectrum is also $\overline{\mathbb{R}^+}$, and the operator has no discrete spectrum.

4. The Schrödinger operator with a constant magnetic field $(B \neq 0)$ in $\mathbb{R}^2$,

$$S_B := \left(D_{x_1} - \frac{Bx_2}{2} \right)^2 + \left(D_{x_2} + \frac{Bx_1}{2} \right)^2, \qquad (10.3.1)$$

with

$$D_{x_j} := \frac{1}{i} \partial_{x_j} = \frac{1}{i} \frac{\partial}{\partial x_j},$$

is another example. The spectrum consists of eigenvalues $(2k + 1)|B|$, but the spectrum is not discrete, because each eigenvalue has infinite multiplicity. This will be treated in the next section.

10.4 On the Schrödinger operator with a constant magnetic field

The Schrödinger operator with a magnetic field was briefly introduced in Section 7.4. In particular, we have given examples where this operator had a compact resolvent. In this section, we analyze in more detail the properties of the Schrödinger operator with a constant magnetic field in dimension 2 and 3. This operator plays an important role in superconductivity theory.

10.4.1 The case of $\mathbb{R}^2$

We analyze the spectrum of

$$S_B := \left(D_{x_1} - \frac{B}{2} x_2 \right)^2 + \left(D_{x_2} + \frac{B}{2} x_1 \right)^2. \qquad (10.4.1)$$

We look first at the self-adjoint realization in $\mathbb{R}^2$. Let us show briefly how its spectrum can be analyzed. We leave it as an exercise to show that the spectra (or the discrete spectra) of two self-adjoint operators S and T are the same if there exists a unitary operator U such that $U(S \pm i)^{-1}U^{-1} = (T \pm i)^{-1}$. We note that this implies that U sends the domain of S onto the domain of T.

In order to determine the spectrum of the operator S_B, we perform a succession of unitary conjugations. The first one is called a gauge transformation. We introduce U_1 on $L^2(\mathbb{R}^2)$, which is defined, for $f \in L^2$, by

$$U_1 f = \exp i B \frac{x_1 x_2}{2} f. \qquad (10.4.2)$$

This satisfies

$$S_B U_1 f = U_1 S_B^1 f, \ \forall f \in \mathcal{S}(\mathbb{R}^2), \tag{10.4.3}$$

with

$$S_B^1 := (D_{x_1})^2 + (D_{x_2} + Bx_1)^2. \tag{10.4.4}$$

Remark 10.6

U_1 is a very special case of what is called a gauge transformation. More generally, we can consider $U = \exp i\phi$, where ϕ is C^∞. If $\Delta_A := \sum_j (D_{x_j} - A_j)^2$ is a general Schrödinger operator associated with a magnetic potential A, then $U^{-1} \Delta_A U = \Delta_{\tilde{A}}$, where $\tilde{A} = A + \text{grad } \phi$. Here we observe that $B := \text{rot } A = \text{rot } \tilde{A}$. The associated magnetic field is unchanged in a gauge transformation. In our example, we are discussing the very special (but important!) case in which the magnetic potential is constant.

We now have to analyze the spectrum of S_B^1.

Observing that the operator has constant coefficients with respect to the variable x_2, we perform a partial Fourier transform with respect to the variable x_2,

$$U_2 = \mathcal{F}_{x_2 \mapsto \xi_2}, \tag{10.4.5}$$

and obtain by conjugation, on $L^2(\mathbb{R}^2_{x_1,\xi_2})$,

$$S_B^2 := (D_{x_1})^2 + (\xi_2 + Bx_1)^2. \tag{10.4.6}$$

We introduce a third unitary transformation, U_3,

$$(U_3 f)(y_1, \xi_2) = f(x_1, \xi_2), \quad \text{with } y_1 = x_1 + \frac{\xi_2}{B}, \tag{10.4.7}$$

and obtain the operator

$$S_B^3 := D_y^2 + B^2 y^2, \tag{10.4.8}$$

operating on $L^2(\mathbb{R}^2_{y,\xi_2})$.

This operator depends only on the variable y. It is easy to find an orthonormal basis of eigenvectors for this operator. We observe in fact that if $f \in L^2(\mathbb{R}_{\xi_2})$ and ϕ_n is the $(n+1)$-th eigenfunction of the harmonic oscillator, then

$$(x, \xi_2) \mapsto |B|^{1/4} f(\xi_2) \cdot \phi_n(|B|^{1/2} y)$$

is an eigenvector corresponding to the eigenvalue $(2n + 1)|B|$. So, each eigenspace has an infinite dimension. An orthonormal basis of this eigenspace can be given by the vectors $e_j(\xi_2)\,|B|^{1/4}\,\phi_n(|B|^{1/2}\,y)$, where e_j $(j \in \mathbb{N})$ is a basis of $L^2(\mathbb{R})$.

Consequently, we have an empty discrete spectrum. The eigenvalues are usually called Landau levels.

10.4.2 The case of $\mathbb{R}^3$

We consider the Schrödinger operator with a constant magnetic field in $\mathbb{R}^3$. After some rotation in $\mathbb{R}^3$, we arrive at the model,

$$P(h, \vec{b}) = D_{x_1}^2 + (D_{x_2} - bx_1)^2 + D_{x_3}^2, \qquad (10.4.9)$$

with

$$b = ||B||.$$

This time, we can take the partial Fourier transform with respect to x_2 and x_3 in order to obtain the operator

$$D_{x_1}^2 + (\xi_2 - bx_1)^2 + \xi_3^2.$$

When $b \neq 0$, we can translate in the variable x_1 and obtain the operator on $L^2(\mathbb{R}^3)$

$$D_{y_1}^2 + (|b|\,y_1)^2 + \xi_3^2.$$

It is then easy to see that the spectrum is $[|b|, +\infty[$. Since no point is isolated, we cannot expect any discrete spectrum.

10.5 Weyl's criterion

We have already mentioned that the essential spectrum is a closed set. In order to determine the essential spectrum, it is useful to have theorems that prove invariance with respect to perturbations. The following characterization is quite useful in this respect.

Theorem 10.7 *Let T be a self-adjoint operator. Then λ belongs to the essential spectrum if and only if there exists a sequence u_n in $D(T)$ with $||u_n|| = 1$ such that u_n tends weakly to 0 and $||(T - \lambda)u_n|| \to 0$ as $n \to \infty$.*

Let us give a proof of this theorem.

The sequence appearing in the theorem is called a Weyl sequence. A point λ such that there exists an associated Weyl sequence is said to belong to the Weyl spectrum $W(T)$. Let us show the inclusion

$$W(T) \subset \sigma_{\text{ess}}(T). \qquad (10.5.1)$$

We have already seen that

$$W(T) \subset \sigma(T). \tag{10.5.2}$$

Suppose, in contradiction, that $\lambda \in \sigma_{\text{disc}}(T)$. Let $\Pi_\lambda := E_{\{\lambda\}}$ be the associated spectral projector. We first observe that since Π_λ has a finite range and hence is compact, we have

$$\Pi_\lambda u_n \rightarrow 0 \in \mathcal{H}. \tag{10.5.3}$$

We define

$$w_n = (I - \Pi_\lambda)u_n,$$

and obtain $\|w_n\| \rightarrow +1$ and $(T - \lambda)w_n = (I - \Pi_\lambda)(T - \lambda)u_n \rightarrow 0$. But $(T - \lambda)$ is invertible on $R(I - \Pi_\lambda)$, so we obtain $w_n \rightarrow 0$ and a contradiction. This shows the announced inclusion (10.5.1).

For the converse, we first observe that if $\lambda \in \sigma_{\text{ess}}(T)$, then, for any $\epsilon > 0$, $\dim R(E_{]\lambda-\epsilon,\lambda+\epsilon[}) = +\infty$. By considering a decreasing sequence ϵ_n such that $\epsilon_n > 0$ and $\lim_{n\rightarrow+\infty} \epsilon_n = 0$, it is easy to obtain an orthonormal system u_n such that $u_n \in R(E_{]\lambda-\epsilon_n,\lambda+\epsilon_n[})$. With this last property, we immediately obtain (see also the proof of Proposition 8.23) the result that $\|(T - \lambda)u_n\| \rightarrow 0$.

Corollary 10.8 *The operator* $-h^2\Delta + V$*, where* V *is a continuous function tending to* 0 *as* $|x| \rightarrow \infty$ *(*$x \in \mathbb{R}^D$*), has* $\overline{\mathbb{R}^+}$ *as its essential spectrum.*

To prove the inclusion of $\overline{\mathbb{R}^+}$ in the essential spectrum, we can consider the sequence

$$u_n(x) = \exp(ix \cdot \xi)n^{-(D/2)} \cdot \chi((x - R_n)/n)$$

with $\chi \geq 0$, supported in the ball $B(0, 1)$ and equal to one on, say, $B(0, \frac{1}{2})$. The sequence R_n is chosen such that $|R_n|$ (for example, $|R_n| = n^2$) tends to ∞ and the supports of the u_n are disjoint.

This is a particular case of a Weyl sequence. The converse can be obtained by abstract analysis and the fact that we know that the essential spectrum of $-\Delta$ is $[0, +\infty[$. This idea will be formalized using the following notion of relative compactness.

Definition 10.9 If T is a closed operator with a dense domain $D(T)$, we say that the operator V is relatively compact with respect to T, or T-compact, if $D(T) \subset D(V)$ and if the image under V of a closed ball in $D(T)$ (for the graph norm $u \mapsto \sqrt{\|u\|^2 + \|Tu\|^2}$) is relatively compact in $\mathcal{H}$.

In other words, we say that V is T-compact if, from each sequence u_n in $D(T)$ that is bounded in $\mathcal{H}$ and such that Tu_n is bounded in $\mathcal{H}$, one can extract a subsequence u_{n_i} such that $V u_{n_i}$ is convergent in $\mathcal{H}$. Here we recall (exercise) that when T is closed, $D(T)$ equipped with the graph norm is a Hilbert space.

Example 10.10 If V is the operator of multiplication by a continuous function V tending to 0, then V is $(-\Delta)$-compact. This is a consequence of Proposition 5.4 and the uniform continuity of V on each compact.

Weyl's theorem states the following.

Theorem 10.11 *Let T be a self-adjoint operator, and let V be symmetric and T-compact. Then $T + V$ is self-adjoint, and the essential spectrum of $T + V$ is equal to the essential spectrum of T.*

The first part of the theorem can be deduced from the discussion in the proof of the Kato–Rellich theorem (Theorem 9.10). We observe, in fact, the following variant of Lions's lemma.

Lemma 10.12 *If V is T-compact and closable, then, for any $a > 0$, there exists $b > 0$ such that*

$$||Vu|| \le a\,||Tu|| + b\,||u||, \quad \forall u \in D(T). \tag{10.5.4}$$

Proof of Lemma 10.12 The proof is by contradiction. If (10.5.4) is not true, then we can find an $a > 0$ such that, $\forall n \in \mathbb{N}^*$, there exist $u_n \in D(T)$ such that

$$a\,||Tu_n|| + n\,||u_n|| < ||Vu_n||. \tag{10.5.5}$$

Observing that $||Vu_n|| \ne 0$ and that the inequality is homogeneous, we can assume in addition that u_n satisfies the condition

$$||Vu_n|| = 1. \tag{10.5.6}$$

From these two properties, we obtain the result that the sequence Tu_n is bounded and that $u_n \to 0$.

On the other hand, by T-compactness, we can extract a subsequence u_{n_k} such that $V u_{n_k}$ is convergent to v with $||v|| = 1$. But $(0, v)$ is in the closure of the graph of V. Since V is closable, this implies $v = 0$ and a contradiction. $\square$

For the second part of Theorem 10.11, we can use Theorem 10.7. If we take a Weyl sequence u_n such that $u_n \to 0$ (weakly) and $(T - \lambda)u_n \to 0$ strongly, we can consider $(T + V - \lambda)u_n$. We have simply to show that we can extract a subsequence u_{n_k} such that $(T + V - \lambda)u_{n_k} \to 0$.

But Tu_n is a bounded sequence. By the T-compactness, we can extract a subsequence such that $V u_{n_k}$ converges strongly to some v in $\mathcal{H}$. It remains to show that $v = 0$. But here we can observe that for any $f \in D(T)$, we have

$$\langle v, f \rangle = \lim_{k \to +\infty} \langle V u_{n_k}, f \rangle = \lim_{k \to +\infty} \langle u_{n_k}, V f \rangle = 0.$$

Here we have used the symmetry of V and the weak convergence of u_n to 0. Using the density of $D(T)$ in $\mathcal{H}$, we obtain $v = 0$. This shows that a Weyl sequence for T is a Weyl sequence for $T + V$.

For the converse, we can intertwine the roles of T and $T + V$, once we have shown that V is $(T + V)$-compact. For this, we can use Lemma 10.12, and observe that the following inequality is true:

$$\|T u\| \leq \frac{1}{1 - a} \left(\|(T + V) u\| + b \|u\| \right). \tag{10.5.7}$$

This shows that if u_n is a sequence such that $(\|u_n\| + \|(T + V) u_n\|)$ is bounded, then this sequence has also the property that $(\|u_n\| + \|T u_n\|)$ is bounded.

10.6 Notes

The application of spectral analysis of the Schrödinger operator with a magnetic field to superconductivity is presented in, for example, [FoHe]. For Weyl's theorem, readers are referred also to [Ro], [HiSi], and [RS-IV]. Weyl sequences are also called Zhislin sequences in [HiSi].

10.7 Exercises

Exercise 10.1 Let V be a C^∞ potential in $\mathbb{R}^2$. Consider, with $B \in \mathbb{R} \setminus \{0\}$, the magnetic Schrödinger operator

$$P = D_{x_1}^2 + (D_{x_2} + B x_1)^2 + V(x),$$

with domain $C_0^\infty(\mathbb{R}^2)$. We assume that V tends to 0 as $|x| \to +\infty$. Determine the essential spectrum of the closure $\bar{P}$ of P.

Exercise 10.2 Let M_f be the operator of multiplication by the real-valued function $f \in C_b^\infty(\mathbb{R}^d)$. Discuss the assumptions under which we can have a discrete or an essential spectrum.

Exercise 10.3 Give an example of a multiplication operator on $\ell^2(\mathbb{Z})$ for which the essential spectrum is not empty.

Exercise 10.4 Consider in $\mathbb{R}^3$ the differential operator (Coulomb Schrödinger operator) $S_0 := -\Delta - 1/r$, initially defined on $C_0^\infty(\mathbb{R}^3)$.

(a) Show that this operator admits a self-adjoint extension S.

(b) Show the continuous injection of $H^2(\mathbb{R}^3)$ into the space of the Hölder functions $C^s(\mathbb{R}^3)$, with $s \in]0, \frac{1}{2}[$, and the compact injection for all compact K of $C^s(K)$ into $C^0(K)$.

(c) Determine the essential spectrum of S. One possibility is to start with an analysis of $S_\chi = -\Delta - \chi/r$, where χ is C^∞ with compact support.

Exercise 10.5 Give natural conditions on the potential V under which the Klein–Gordon operator $\sqrt{1 - \Delta} + V$ is essentially self-adjoint, starting from $\mathcal{S}(\mathbb{R}^3)$.

11

The max–min principle

The max–min principle provides an alternative way to describe the lowest part of the spectrum when it is discrete. It also gives an efficient way to localize these eigenvalues and to follow their dependence on various parameters.

11.1 On nonnegativity

We first recall the following definition.

Definition 11.1 Let A be a symmetric operator. We say that A is nonnegative (and write $A \geq 0$) if

$$\langle Au, u \rangle \geq 0, \ \forall u \in D(A). \tag{11.1.1}$$

The following proposition relates the positivity to the spectrum.

Proposition 11.2 *Let A be a self-adjoint operator. Then $A \geq 0$ if and only if $\sigma(A) \subset [0, +\infty[$.*

Proof It is clear that if the spectrum is in $\mathbb{R}^+$, then the operator is nonnegative. This can be seen, for example, using the spectral representation:

$$\langle Au, u \rangle = \int_{\lambda \in \sigma(A)} \lambda \, d\|E_\lambda u\|^2.$$

Now, if $A \geq 0$, then for any $a > 0$, $A + a$ is invertible. We have

$$a \, \|u\|^2 \leq \langle (A + a)u, u \rangle \leq \|(A + a)u\| \, \|u\|,$$

which leads to

$$a \, \|u\| \leq \|(A + a)u\|, \ \forall u \in D(A). \tag{11.1.2}$$

This inequality gives the closed range and the injectivity. Since A is self-adjoint, we also obtain from the injectivity the result that the range of $(A + a)$ is dense. This shows that $-a$ is not in the spectrum of A. $\qquad\qquad\square$

Example 11.3 Consider the Schrödinger operator $P := -\Delta + V$, with $V \in C^\infty$ and semibounded. Then

$$\sigma(P) \subset [\inf V, +\infty[\,. \qquad\qquad (11.1.3)$$

11.2 Variational characterization of the discrete spectrum

Theorem 11.4 *Let A be a self-adjoint semibounded operator. Let $\Sigma := \inf \sigma_{\text{ess}}(A)$, and consider $\sigma(A) \cap]-\infty, \Sigma[$, described as a sequence (finite or infinite) of eigenvalues $\lambda_j(A)$ (or, more simply, λ_j), which is ordered in increasing order:*

$$\lambda_1 < \lambda_2 < \cdots < \lambda_n \cdots.$$

Then we have

$$\lambda_1 = \inf_{\phi \in D(A), \phi \neq 0} ||\phi||^{-2} \langle A\phi, \phi\rangle, \qquad\qquad (11.2.1)$$

$$\lambda_2 = \inf_{\phi \in D(A) \cap K_1^\perp, \phi \neq 0} ||\phi||^{-2} \langle A\phi, \phi\rangle, \qquad\qquad (11.2.2)$$

and, for $n \geq 2$,

$$\lambda_n = \inf_{\phi \in D(A) \cap K_{n-1}^\perp, \phi \neq 0} ||\phi||^{-2} \langle A\phi, \phi\rangle, \qquad\qquad (11.2.3)$$

where

$$K_j = \oplus_{i \leq j} N(A - \lambda_i).$$

Proof

Step 1. We start with the lowest eigenvalue. We define $\widehat{\mu}_1(A)$ by

$$\widehat{\mu}_1(A) := \inf_{\phi \in D(A), \phi \neq 0} ||\phi||^{-2} \langle A\phi, \phi\rangle. \qquad\qquad (11.2.4)$$

If ϕ_1 is an eigenfunction associated with λ_1, we immediately obtain the inequality

$$\widehat{\mu}_1(A) \leq \lambda_1(A). \qquad\qquad (11.2.5)$$

Let us now prove the converse inequality. Using the spectral theorem, we immediately obtain the result that $A \geq \inf \sigma(A)$. So, we obtain

$$\inf \sigma(A) \le \widehat{\mu}_1(A). \tag{11.2.6}$$

Now, if the spectrum below Σ is not empty, we obtain

$$\lambda_1(A) \le \widehat{\mu}_1(A).$$

Consequently, we have the equality, but we can actually obtain a little more. In fact, we have proved that if $\widehat{\mu}_1(A) < \Sigma$, then the spectrum below Σ is not empty. In particular, the bottom of the spectrum is an eigenvalue equal to $\widehat{\mu}_1(A)$.

Step 2. The proof is by recursion, applying Step 1 to $A_{/D(A) \cap K_{n-1}^{\perp}}$.

This ends the proof of Theorem 11.4. $\qquad\qquad\qquad\qquad\qquad\qquad \square$

Example 11.5 (Payne–Polya–Weinberger inequality.) Let $P = -\Delta + V$, with $V \in C^{\infty}$ nonnegative and $V \to +\infty$ as $|x| \to +\infty$. We assume that V is even, i.e.,

$$V(x) = V(-x). \tag{11.2.7}$$

Then λ_2 satisfies

$$\lambda_2 \le \inf_{\phi \in Q(P), \phi \text{ odd and } ||\phi|| = 1} \langle P\phi, \phi \rangle. \tag{11.2.8}$$

Let u_1 be the first normalized eigenfunction. We assume that the lowest eigenvalue of the Schrödinger operator is simple (this is a variant of the Krein–Rutman theorem) and that the first eigenfunction can be chosen positive, with exponential decay at ∞ together with ∇u_1 (this is a consequence of Agmon's inequality). It is not difficult to verify that u_1 is even. Let us consider $v_j := x_j u_1$; we know that v_j is in the form domain of P. We observe that

$$P(x_j u_1) = \lambda_1 x_j u_1 - 2\partial_{x_j} u_1.$$

Taking the scalar product with $x_j u_1$, we then obtain

$$\begin{aligned}(\lambda_2 - \lambda_1) \, ||x_j u_1||^2 &\le -2\langle \partial_{x_j} u_1, \, x_j u_1 \rangle \\ &\le ||u_1||^2 \\ &\le 1.\end{aligned} \tag{11.2.9}$$

We now use the uncertainty principle (1.3.13) and obtain

$$(\lambda_2 - \lambda_1) \le 4 \, ||\partial_{x_j} u_1||^2. \tag{11.2.10}$$

On the other hand,

$$||\nabla u_1||^2 + \int_{\mathbb{R}^m} V(x) |u_1(x)|^2 \, dx_1 \ldots dx_m = \lambda_1, \qquad (11.2.11)$$

and this gives

$$||\nabla u_1||^2 \leq \lambda_1. \qquad (11.2.12)$$

Putting the inequalities (11.2.9) and (11.2.12) together and summing over j, we obtain

$$\lambda_2 - \lambda_1 \leq \frac{4}{m}\lambda_1. \qquad (11.2.13)$$

This inequality is not optimal, in the sense that for $m = 1$ and $V(x) = x^2$, we have $\lambda_2 - \lambda_1 = 2\lambda_1$.

Example 11.6 Consider $S_h := -h^2 \Delta + V$ on $\mathbb{R}^m$, where V is a C^∞ potential tending to 0 at ∞ and such that $\inf_{x \in \mathbb{R}^m} V(x) < 0$.

Then, if $h > 0$ is small enough, there exists at least one eigenvalue for S_h. We note that the essential spectrum is $[0, +\infty[$. The proof of the existence of this eigenvalue is elementary. If $x_{\min}$ is a point such that $V(x_{\min}) = \inf_x V(x)$, it is enough to show that, with $\phi_h(x) = \exp -(\lambda/h) |x - x_{\min}|^2$, the quotient $\langle S_h \phi_h, \phi_h \rangle / ||\phi_h||^2$ tends as $h \to 0$ to $V(x_{\min})$, which is negative by assumption.

11.3 Max–min principle

We now give a more flexible criterion for the determination of the bottom of the spectrum and of the essential spectrum. This flexibility comes from the fact that we do not need an explicit knowledge of the various eigenspaces.

Theorem 11.7 *Let $\mathcal{H}$ be a Hilbert space of infinite dimension,[1] and let A be a self-adjoint semibounded operator of domain $D(A) \subset \mathcal{H}$. We introduce*

$$\mu_n(A) = \sup_{\psi_1, \psi_2, \ldots, \psi_{n-1}} \quad \inf_{\left\{ \begin{array}{l} \phi \in [\text{span}(\psi_1, \ldots, \psi_{n-1})]^\perp; \\ \phi \in D(A) \text{ and } ||\phi|| = 1 \end{array} \right\}} \langle A\phi, \phi \rangle_{\mathcal{H}}. \qquad (11.3.1)$$

Then either

(a) *$\mu_n(A)$ is the n-th eigenvalue when the eigenvalues are ordered in increasing order (counting the multiplicity) and A has a discrete spectrum in $] -\infty, \mu_n(A)]$, or*

---

[1] In the case of a finite-dimensional Hilbert space of dimension d, the minimax principle holds for $n \leq d$.

(b) $\mu_n(A)$ *corresponds to the bottom of the essential spectrum. In this case, we have* $\mu_j(A) = \mu_n(A)$ *for all* $j \geq n$.

Remark 11.8 When the operator has a compact resolvent, case (b) does not occur and the supremum in (11.3.1) is a maximum. Similarly, the infimum is a minimum. This explains the traditional terminology "max–min principle" for this theorem.

Proof If Ω is a Borelian set, let E_Ω be the projection-valued measure for A. We first prove that

$$\dim R\left(E_{]-\infty,a[}\right) < n \text{ if } a < \mu_n(A), \tag{11.3.2}$$

$$\dim R(E_{]-\infty,a[}) \geq n \text{ if } a > \mu_n(A). \tag{11.3.3}$$

Note that the conjunction of (11.3.2) and (11.3.3) shows that $\mu_n(A)$ is in the spectrum of A.

Step 1: Proof of (11.3.2).

Let a and n be given such that $a < \mu_n(A)$. We prove (11.3.2) by contradiction. If this equation was false, then we would have $\dim R(E_{]-\infty,a[})) \geq n$, and we would be able to find an n-dimensional space $V \subset R(E_{]-\infty,a[})$. Note now that since A is bounded from below, $R(E_{]-\infty,a[})$ is included in $D(A)$.

So, we can find an n-dimensional space $V \subset D(A)$ such that

$$\forall \phi \in V, \ \langle A\phi, \phi \rangle \leq a \, ||\phi||^2. \tag{11.3.4}$$

But then, given any $\psi_1, \cdots, \psi_{n-1}$ in $\mathcal{H}$, we can find $\phi \in V \cap \{\psi_1, \cdots, \psi_{n-1}\}^\perp$ such that $||\phi|| = 1$ and $\langle A\phi, \phi \rangle \leq a$. Returning to the definition, this shows that $\mu_n(A) \leq a$, and we have a contradiction.

Note that we have proved the following proposition in this step.

Proposition 11.9 *Suppose that there exists an a and an n-dimensional subspace $V \subset D(A)$ such that (11.3.4) is satisfied. Then we have the following inequality:*

$$\mu_n(A) \leq a. \tag{11.3.5}$$

Modulo the complete proof of the theorem, we obtain the following corollary.

Corollary 11.10 *Under the same assumptions as in Proposition 11.9, if a is below the bottom of the essential spectrum of A, then A has at least n eigenvalues (counted with multiplicity).*

Step 2: Proof of (11.3.3).

Suppose that (11.3.3) is false. Then $a > \mu_n(A)$ and dim $R(E_{]-\infty,a[}) \leq n - 1$, so we can find $(n - 1)$ generators $\psi_1, \cdots, \psi_{n-1}$ of this space. Then any $\phi \in D(A) \cap \text{span}\{\psi_1, \cdots, \psi_{n-1}\}^\perp$ is in $R(E_{[a,+\infty[})$, so

$$\langle A\phi, \; \phi \rangle \geq a \, ||\phi||^2.$$

Therefore, returning to the definition of $\mu_n(A)$, we obtain $\mu_n(A) \geq a$, in contradiction to our initial assumption.

Before we continue the proof, let us emphasize one point.

Remark 11.11 In the definition of $\mu_n(A)$, the $\psi_1, \cdots, \psi_{n-1}$ are assumed only to belong to the Hilbert space $\mathcal{H}$.

Step 3: $\mu_n(A) < +\infty$.

First, the semiboundedness of A from below gives a uniform lower bound.

Secondly, if $\mu_n(A) = +\infty$, this would mean, by (11.3.2), that dim $R(E_{]-\infty,a[}) < n$ for all a and, consequently, that $\mathcal{H}$ is finite dimensional. This is a contradiction if $\mathcal{H}$ is infinite dimensional. But the finite case is trivial; in fact, we have $\mu_n(A) \leq ||A||$ in this case.

As the statement of the theorem suggests, there are two cases to consider, and this will be the object of the next two steps.

Step 4. We first assume (with $\mu_n = \mu_n(A)$) that

$$\dim R(E_{]-\infty,\mu_n+\epsilon[}) = +\infty, \; \forall \epsilon > 0. \tag{11.3.6}$$

We claim that the second alternative in the theorem applies in this case. Using (11.3.2) and (11.3.6), we obtain

$$\dim R(E_{]\mu_n-\epsilon,\mu_n+\epsilon[}) = +\infty, \; \forall \epsilon > 0. \tag{11.3.7}$$

This shows that $\mu_n(A) \in \sigma_{\text{ess}}(A)$.

On the other hand, using (11.3.2) again, we immediately obtain the result that $]-\infty, \mu_n(A)[$ does not contain any point in the essential spectrum. Thus $\mu_n(A) = \inf\{\lambda \mid \lambda \in \sigma_{\text{ess}}(A)\}$.

Let us now show that $\mu_{n+1} = \mu_n$ in this case. From the definition of the $\mu_k(A)$, it is clear that $\mu_{n+1} \geq \mu_n$, since we can take $\psi_n = \psi_{n-1}$. But if $\mu_{n+1} > \mu_n$, (11.3.2) would also be satisfied for μ_{n+1}, and this is in contradiction with (11.3.6).

Step 5. We now assume that

$$\dim R(E_{]-\infty,\mu_n+\epsilon_0[}) < \infty, \quad \text{for some } \epsilon_0 > 0. \tag{11.3.8}$$

It is clear that the spectrum is discrete in $]-\infty, \mu_n + \epsilon_0[$. Then, for $\epsilon_1 > 0$ small enough,

$$R(E_{]-\infty,\mu_n]}) = R(E_{]-\infty,\mu_n+\epsilon_1[}),$$

and by (11.3.3),

$$\dim R(E_{]-\infty,\mu_n]}) \geq n. \tag{11.3.9}$$

So, there are at least n eigenvalues $E_1 \leq E_2 \leq \cdots \leq E_n \leq \mu_n$ for A. If E_n were strictly less than μ_n, then dim (Range $E_{]-\infty,E_n]}$) would equal n, in contradiction with (11.3.2). Therefore $E_n = \mu_n$, and μ_n is an eigenvalue.

This ends the proof of Theorem 11.7. $\qquad\square$

The following theorem is a natural extension of Theorem 11.7.

Theorem 11.12 *Let A be a self-adjoint semibounded operator and let $Q(A)$ be its form domain.*[2] *Then*

$$\mu_n(A) = \sup_{\psi_1,\psi_2,\dots,\psi_{n-1}} \quad \inf_{\left\{ \begin{array}{l} \phi \in [\mathrm{span}(\psi_1,\dots,\psi_{n-1})]^\perp; \\ \phi \in Q(A) \text{ and } \|\phi\|=1 \end{array} \right\}} \langle A\phi, \ \phi \rangle_{\mathcal{H}}. \tag{11.3.10}$$

Proof Let $\tilde{\mu}_n$ be the right-hand side of (11.3.10). By imitating the proof of Theorem 11.7, we obtain the result that each $\tilde{\mu}_n$ obeys one of the two conditions in that theorem. These conditions determine μ_n, and, consequently, $\mu_n = \tilde{\mu}_n$. $\qquad\square$

One may also note (see Section 3.3) that when we constructed the Friedrichs extension, we showed that the domain of the Friedrichs extension is dense in the form domain.

Applications

- It is often very useful to apply the max–min principle by taking the minimum over a dense set in the form domain $Q(A)$.
- The max–min principle permits one to check the continuity of the eigenvalues with respect to parameters. For example, the lowest eigenvalue $\lambda_1(\epsilon)$

---

[2] This is associated, by completion, with the quadratic form $u \mapsto \langle u, Au \rangle_{\mathcal{H}}$, initially defined on $D(A)$.

of $-d^2/dx^2 + x^2 + \epsilon x^4$ is increasing with respect to ϵ. It can be shown that $\epsilon \mapsto \lambda_1(\epsilon)$ is right continuous on $[0, +\infty[$. (The reader can assume that the corresponding eigenfunction is in $\mathcal{S}(\mathbb{R})$ for $\epsilon \geq 0$).

- The max–min principle permits one to give an upper bound on the bottom of the spectrum and to compare the spectra of two operators. If $A \leq B$ in the sense that $Q(B) \subset Q(A)$ and[3]

$$\langle Au, u \rangle \leq \langle Bu, u \rangle, \ \forall u \in Q(B),$$

then

$$\lambda_n(A) \leq \lambda_n(B).$$

Similar conclusions occur if we have $D(B) \subset D(A)$.

Example 11.13 (Comparison of Dirichlet and Neumann realizations.) Let Ω be a bounded regular connected open set in $\mathbb{R}^m$. Then the N-th eigenvalue of the Neumann realization of $-\Delta + V$ is less than or equal to the N-th eigenvalue of the Dirichlet realization. It is in fact enough to observe the inclusion of the form domains.

Example 11.14 (Monotonicity with respect to the domain.) Let $\Omega_1 \subset \Omega_2 \subset \mathbb{R}^m$ be two bounded regular open sets. Then the n-th eigenvalue of the Dirichlet realization of the Schrödinger operator in Ω_2 is less than or equal to the n-th eigenvalue of the Dirichlet realization of the Schrödinger operator in Ω_1. We observe that we can identify $H_0^1(\Omega_1)$ with a subspace of $H_0^1(\Omega_2)$ by considering just an extension by 0 in $\Omega_2 \setminus \Omega_1$.

We then have

$$
\begin{aligned}
\lambda_n(\Omega_2) &= \sup_{\{\psi_1, \cdots, \psi_{n-1} \in L^2(\Omega_2)\}} \inf_{\left\{ \substack{\phi \in H_0^1(\Omega_2) \\ \langle \phi, \psi_j \rangle_{L^2(\Omega_2)} \text{ and } \|\phi\|=1} \right\}} \|\nabla\phi\|^2_{L^2(\Omega_2)} \\
&\leq \sup_{\{\psi_1, \cdots, \psi_{n-1} \in L^2(\Omega_2)\}} \inf_{\left\{ \substack{\phi \in H_0^1(\Omega_1) \\ \langle \phi, \psi_j \rangle_{L^2(\Omega_2)} \text{ and } \|\phi\|=1} \right\}} \|\nabla\phi\|^2_{L^2(\Omega_2)} \\
&= \sup_{\{\psi_1, \cdots, \psi_{n-1} \in L^2(\Omega_2)\}} \inf_{\left\{ \substack{\phi \in H_0^1(\Omega_1) \\ \langle \phi, \psi_j \rangle_{L^2(\Omega_1)} \text{ and } \|\phi\|=1} \right\}} \|\nabla\phi\|^2_{L^2(\Omega_1)} \\
&= \sup_{\{\psi_1, \cdots, \psi_{n-1} \in L^2(\Omega_1)\}} \inf_{\left\{ \substack{\phi \in H_0^1(\Omega_1) \\ \langle \phi, \psi_j \rangle_{L^2(\Omega_1)} \text{ and } \|\phi\|=1} \right\}} \|\nabla\phi\|^2_{L^2(\Omega_1)} \\
&= \lambda_n(\Omega_1).
\end{aligned}
$$

Note that this argument is not valid for the Neumann realization.

---

[3] It is enough to verify the inequality on a dense set in $Q(B)$.

Example 11.15 (Courant's nodal theorem.) Courant's nodal theorem states that when we consider the Dirichlet realization of the Laplacian in a bounded regular connected set $\Omega \subset \mathbb{R}^m$, if λ_j denotes the (increasing) sequence of the eigenvalues counted with multiplicity and u_k denotes a real eigenfunction associated with λ_k, then $\{x \in \Omega \,|\, u_k(x) \neq 0\}$ has at most k components.

Let us sketch a proof of this theorem, which involves the techniques of this chapter. Suppose $\lambda_{k-1} < \lambda_k$. If $u_k \in E(\lambda_k)$ has more than k nodal components, then we can construct a real vector space of dimension k with energy less than λ_k and with functions vanishing in some fixed open set (a component of the zero set of u_k). In addition, we can find in this space a nonzero function that is orthogonal to the first $(k-1)$ eigenfunctions. Then this function should be an eigenfunction corresponding to λ_k and vanish in an open set. In addition, a standard result in the field of partial differential equations states that eigenfunctions are analytic. Hence this function is identically 0 by analyticity, and we have a contradiction.

Note also that any real second eigenfunction has exactly two nodal sets. It has more than one because it has to be orthogonal to the first eigenfunction, which is strictly positive (by the Krein–Rutman theorem), and Courant's theorem states that it has at most two.

11.4 The Cwickel–Lieb–Thirring inequality

In order to complete the picture, let us mention that if $m \geq 3$, then the following theorem, due to Cwickel, Lieb, and Rozenbljum (CLR), holds.

Theorem 11.16 *There exists a constant L_m such that for any V such that $V_- \in L^{m/2}$, and if $m \geq 3$, the number N_- of negative eigenvalues of S_1 is finite and bounded by*

$$N_- \leq L_m \int_{V(x) \leq 0} (-V)^{m/2} dx. \tag{11.4.1}$$

This shows that when $m \geq 3$, there are examples of negative potentials V (which are not identically zero) such that the corresponding Schrödinger operator S_1 has no eigenvalues. A sufficient condition is

$$L_m \int_{V \leq 0} (-V)^{m/2} \, dx < 1.$$

In the other direction, we have the following result.

Proposition 11.17 *Let V be in $C_0^\infty(\mathbb{R}^m)$ $(m = 1, 2)$. If*

$$\int_{\mathbb{R}^m} V(x)\,dx < 0, \tag{11.4.2}$$

then the Schrödinger operator has at least one negative eigenvalue.

Proof Here, we treat only the case where $V \in C_0^\infty(\mathbb{R}^m)$.

We first observe that the essential spectrum is $[0, +\infty[$. For the proof of the proposition, it is then enough to find $\psi \in D(S_1)$ such that

$$\langle S_1 \psi, \ \psi \rangle_{L^2(\mathbb{R}^m)} < 0.$$

When $m = 1$, introducing the trial function $\psi_a(x) = \exp -a\,|x|$ for $a > 0$, we obtain

$$\int_{\mathbb{R}} |\psi_a'(x)|^2\,dx = a,$$

and

$$\lim_{a \to 0} \int_{\mathbb{R}} V(x)\,|\psi_a(x)|^2\,dx = \int_{\mathbb{R}} V(x)\,dx < 0,$$

by the dominated convergence theorem. Hence the sum becomes negative for a small enough.

When $m = 2$, we can take $\psi_a(x) = \exp -\frac{1}{2}|x|^a$, $a > 0$, and then

$$\int_{\mathbb{R}^2} ||\nabla \psi_a(x)||^2\,dx = \frac{\pi}{2}a$$

and

$$\lim_{a \to 0} \int_{\mathbb{R}^2} V(x)\,|\psi_a(x)|^2\,dx = e^{-1/2} \int_{\mathbb{R}^2} V(x)\,dx < 0. \qquad \square$$

11.5 Essential spectrum and Persson's theorem

Here, we mention some results due to S. Agmon.

Theorem 11.18 *Let V be a real-valued potential satisfying the conditions of Theorem 9.10 (with $A = -\Delta$ and $B = V$), and let $H = -\Delta + V$ be the corresponding self-adjoint semibounded Schrödinger operator with domain $H^2(\mathbb{R}^m)$. Then the bottom of the essential spectrum is given by*

$$\inf \sigma_{\text{ess}}(H) = \Sigma(H), \tag{11.5.1}$$

where

$$\Sigma(H) := \sup_{\mathcal{K} \subset \mathbb{R}^m} \left[\inf_{||\phi||=1} \{ \langle \phi, H\phi \rangle \mid \phi \in C_0^\infty(\mathbb{R}^m \setminus \mathcal{K}) \} \right], \qquad (11.5.2)$$

and where the supremum is over all compact subsets $\mathcal{K} \subset \mathbb{R}^m$.

Essentially, this is a corollary of Weyl's theorem (Theorem 10.11). The following property can be used to prove it.

Lemma 11.19

$$\sigma_{\text{ess}}(H) = \sigma_{\text{ess}}(H + W),$$

for any regular potential W with compact support.

11.6 Notes

Courant's theorem is proven in [CH]. Agmon's estimates are proven in [Ag]. For Persson's theorem, readers are referred to Agmon's book [Ag] for details. For the Cwickel–Lieb–Rozenblyum theorem, readers are referred to [RS-IV], p. 101. The counterexamples presented in the proof of Proposition 11.17 reduce when $m = 1$ to those presented by Avron, Herbst, and Simon [AHS], and when $m = 2$ to those presented by Blanchard and Stubbe [BS]. These CLR inequalities are just one example of many celebrated inequalities, including also the so-called Lieb–Thirring inequalities and Berezin–Li–Yau inequalities (see Exercise 11.9 below and [Lap, LiYa]).

11.7 Exercises

Exercise 11.1 (Temple's inequality.) Let A be a self-adjoint operator on a Hilbert space, and let $\psi \in D(A)$ such that $||\psi|| = 1$. Suppose that $\sigma(A) \cap \,]\alpha, \beta[\, = \{\lambda\}$ in some interval $]\alpha, \beta[$, and that $\eta = \langle \psi, A\psi \rangle$ belongs to the interval $]\alpha, \beta[$. Show that

$$\eta - \frac{\epsilon^2}{\beta - \eta} \le \lambda \le \eta + \frac{\epsilon^2}{\eta - \alpha},$$

where

$$\epsilon^2 = ||(A - \eta)\psi||^2.$$

As a preliminary result, we suggest that you show that $(A - \alpha)(A - \lambda)$ and $(A - \beta)(A - \lambda)$ are nonnegative operators. Then apply the inequalities to ψ.

Show that the above inequality is an improvement if $\epsilon^2 \le (\beta - \eta)(\eta - \alpha)$.

Compare this with what is given by the spectral theorem or the max–min principle.

Exercise 11.2 In continuation of Example 11.6, show that for any $\epsilon > 0$ and any N, there exists $h_0 > 0$ such that for $h \in]0, h_0]$, S_h has at least N eigenvalues in $[\inf V, \inf V + \epsilon]$. We suggest that you treat first the case in which V has a unique nondegenerate minimum at 0.

Exercise 11.3 Consider the Neumann realization in $\mathbb{R}^+$ of $P_0(\xi) :=$ $D_t^2 + (t - \xi)^2$, where ξ is a parameter in $\mathbb{R}$. Find an upper bound for $\Theta_0 = \inf_\xi \mu(\xi)$, where $\mu(\xi)$ is the smallest eigenvalue of $P_0(\xi)$. Following the physicist Charles Kittel, we suggest that you proceed by minimizing $\langle P_0(\xi)\phi(\cdot; \rho), \phi(\cdot; \rho)\rangle$ over the normalized functions $\phi(t; \rho) := c_\rho \exp -\rho t^2$ ($\rho > 0$). For which value of ξ is this quantity minimal?

Deduce the inequality

$$\Theta_0 < \sqrt{1 - \frac{2}{\pi}}.$$

Exercise 11.4 In the same spirit as in Exercise 11.3, find an upper bound for the lowest eigenvalue of the quartic operator $D_t^2 + \frac{1}{4}t^4$.

Using a comparison with the harmonic oscillator $D_t^2 + \alpha t^2 + \beta$, find a lower bound that is optimal with respect to the method.

Using a comparison with $D_t^2 + V_\alpha(t)$, where $V_\alpha(t) = 0$ for $|t| \leq \alpha$ and $V_\alpha(t) = \frac{1}{4}\alpha^4$ for $|t| \geq \alpha$, and by optimizing over α, find an alternative lower bound, and compute it with the help of a computer.

Exercise 11.5 Let H_a be the Dirichlet realization of $-d^2/dx^2 + x^2$ in $]-a, +a[$. Show that the lowest eigenvalue $\lambda_1(a)$ of H_a is positive and monotonically decreasing as $a \to +\infty$, and tends exponentially fast to 1 as $a \to +\infty$. Give an estimate of $|\lambda_1(a) - 1|$ that is as accurate as possible.

Exercise 11.6 Let[4] Ω be a bounded regular set in $\mathbb{R}^d$. Denote by λ_n ($n \geq 1$) and μ_n the sequences of eigenvalues of the Laplacian for the Dirichlet and Neumann problems, respectively.

(a) Show that

$$\mu_{n+1} \leq \lambda_n, \ \forall n \geq 1.$$

We suggest that you use the minimax principle and analyze the quantity $\int_\Omega |\nabla u|^2 \, dx$ for

$$u = \sum_{j=1}^{n} \alpha_j \phi_j^D(x) + \beta e^{i\xi \cdot x},$$

with $|\xi|^2 = \lambda_n$.

---

[4] This was inspired by a question in a paper by Benguria, Levitin, and Parnovski [BLP] and by discussions with A. Laptev.

(b) Is the inequality strict?

(c) Now consider the function $\xi \mapsto \chi(\xi) = \int_\Omega e^{i\xi \cdot x} \, dx$. Show that if Ω is balanced (that is, if $x \in \Omega$, then $-x \in \Omega$), then χ is real.

(d) Denote by $\kappa(\Omega)$ the function

$$\kappa(\Omega) = \inf\{\xi \in \mathbb{R}^d, \ \chi(\xi) = 0\}.$$

Show that if $\sqrt{\lambda_n(\Omega)} > 2\kappa(\Omega)$, then $\mu_{n+2}(\Omega) \leq \lambda_n(\Omega)$. This time, we suggest that you use the minimax principle and analyze the quantity $\int_\Omega |\nabla u|^2 \, dx$ for $u = \sum_{j=1}^n \alpha_j \phi_j^D(x) + \beta_1 e^{i\xi_1 \cdot x} + \beta_2 e^{i\xi_2 \cdot x}$, with $|\xi_1|^2 = |\xi_2|^2 = \lambda_n$ and $\chi(\xi_1 - \xi_2) = 0$.

Exercise 11.7 Consider in $\mathbb{R}^3$ the differential operator $S_0 := -\Delta - 1/r$, initially defined on $C_0^\infty(\mathbb{R}^3)$, which we have discussed in Exercise 10.4, and let S be its self-adjoint extension.

(a) Show, using the minimax principle, that S has at least one negative eigenvalue. We suggest that you try to minimize over $au \mapsto \langle S_0 u, u \rangle / \|u\|^2$ with $u(x) = \exp -ar$.

(b) Determine this lowest eigenvalue (using the property that the ground state should be radial).

Exercise 11.8 Consider in $L^2(\mathbb{R}^2)$ the magnetic Schrödinger operator

$$P_0 := D_x^2 + \left(D_y - \frac{x}{x^2 + y^2 + 1} \right)^2.$$

Show that it has a natural self-adjoint extension, and determine its essential spectrum.

Exercise 11.9 (Berezin–Li–Yau inequalities.) If Ω is a bounded regular open set in $\mathbb{R}^m$ and $H(\Omega)$ denotes the Dirichlet realization of the Laplacian in Ω, show that for any $\lambda > 0$,

$$\sum_k (\lambda - \lambda_k)_+ \leq (2\pi)^{-m} |\Omega| \int (\lambda - \xi^2)_+ \, d\xi,$$

where $|\Omega|$ is the volume of Ω.

We suggest that you start by showing the following formulas:

$$\lambda_k = \int |\xi|^2 \, \widehat{|\tilde\phi_k|}^2 \, d\xi,$$

where $\tilde\phi_k$ is the extension by 0 outside of Ω of the normalized eigenfunction ϕ_k associated with λ_k; and, using Plancherel's formula (2.1.7),

$$\sum_k \widehat{|\tilde\phi_k|}^2 = (2\pi)^{-d} |\Omega|.$$

12

Spectral questions about the Rayleigh equation

Fluid mechanics (in particular, the analysis of the stability of solutions) provides a lot of questions in spectral theory. The aim of this chapter is to describe one of these questions and its solution, which turns out in the end to be simple (using the material previously introduced).

12.1 The physical motivation behind the problem

The starting point in fluid mechanics is the analysis of the following differential systems in $\mathbb{R}^4 = \mathbb{R}^3_x \times \mathbb{R}_t$:

$$\begin{aligned} \partial_t \rho + \operatorname{div}(\rho \vec{u}) &= 0, \\ \partial_t(\rho \vec{u}) + \rho(\vec{u} \cdot \nabla)\vec{u} + \nabla p &= \rho \vec{g}. \end{aligned} \tag{12.1.1}$$

The unknowns are the velocity vector $\vec{u} = (u_1, u_2, u_3)$, the density ρ, and the pressure p. We assume that the gravity field is vertical, i.e., $\vec{g} = (0, 0, 1)g$. More explicitly, the first line reads

$$\partial_t \rho + \sum_{j=1}^{3} \partial_{x_j}(\rho u_j) = 0,$$

and the second line is the system of three equations ($i = 1, 2, 3$)

$$\partial_t(\rho u_i) + \rho\left(\sum_j u_j \cdot \partial_{x_j} u_i\right) + \partial_i p = \rho g \, \delta_{i3}.$$

This system models the so-called Rayleigh–Taylor instability, which occurs when a heavy fluid is located above a light fluid in a gravity field directed from the heavy to the light fluid. The question is to study the linear growth rate of this instability in a situation where there is a mixing region. This linear growth rate corresponds to γ in (12.1.15) below.

Our aim is to analyze the linearized problem around the solution

$$\rho = \rho^0, \ u = u^0 = 0, \ p = p^0, \tag{12.1.2}$$

where ρ^0 is assumed to depend only on x_3, and p^0 and ρ^0 are related by

$$\nabla p^0 = \rho^0 \vec{g}. \tag{12.1.3}$$

This solution is a stationary (i.e., time-independent) solution of the system (12.1.1). We assume that the perturbation $(\hat{u}, \hat{p}, \hat{\rho})$ is incompressible; that is, it satisfies

$$\operatorname{div} \vec{\hat{u}} = 0. \tag{12.1.4}$$

The linearized system takes the form

$$\partial_t \hat{\rho} + (\rho^0)' \hat{u}_3 = 0, \tag{12.1.5}$$

$$\rho^0 \partial_t \hat{u}_1 + \partial_{x_1} \hat{p} = 0, \tag{12.1.6}$$

$$\rho^0 \partial_t \hat{u}_2 + \partial_{x_2} \hat{p} = 0, \tag{12.1.7}$$

$$\rho^0 \partial_t \hat{u}_3 + \partial_{x_3} \hat{p} = g \hat{\rho}. \tag{12.1.8}$$

In order to analyze this system (at least formally), we form an equation for $\hat{u}_3$ (by eliminating the other unknowns). We first differentiate with respect to t (12.1.8). This leads to

$$\rho^0 \partial_t^2 \hat{u}_3 + \partial_{tx_3}^2 \hat{p} = g \frac{\partial \hat{\rho}}{\partial t}. \tag{12.1.9}$$

We use (12.1.5) in order to eliminate $\partial \hat{\rho}/\partial t$, and obtain

$$\rho^0 \partial_t^2 \hat{u}_3 + \partial_{tx_3}^2 \hat{p} + g(\rho^0)'(x_3)\hat{u}_3 = 0. \tag{12.1.10}$$

We now differentiate (12.1.6) and (12.1.7) with respect to x_1 and x_2, respectively. This gives

$$\rho^0 \partial_{tx_1}^2 \hat{u}_1 + \partial_{x_1}^2 \hat{p} = 0 \tag{12.1.11}$$

and

$$\rho^0 \partial^2_{tx_2} \hat{u}_2 + \partial^2_{x_2} \hat{p} = 0. \tag{12.1.12}$$

Differentiating (12.1.4) with respect to t and using (12.1.11) and (12.1.12), we obtain

$$\Delta_{1,2} \hat{p} = \rho^0 \partial^2_{tx_3} \hat{u}_3. \tag{12.1.13}$$

It remains to eliminate $\hat{p}$ between (12.1.10) and (12.1.13):

$$\Delta_{12}(\rho^0 \partial^2_t \hat{u}_3 + (\rho^0)'g\hat{u}_3) + (\rho^0)'\partial^3_{x_3tt}\hat{u}_3 = 0. \tag{12.1.14}$$

We first look for a solution $\hat{u}_3$ in the form

$$\hat{u}_3(x_1, x_2, x_3t; k_1, k_2) = v(x_3) \exp(\gamma t + ik_1x_1 + ik_2x_2). \tag{12.1.15}$$

The next step, which will not be described here, is to look for a superposition of such solutions in the form $\int \theta(k_1, k_2)\hat{u}_3(x_1, x_2, x_3; t, k_1, k_2) \, dk_1 \, dk_2$. This leads to the ordinary differential equation

$$-(k_1^2 + k_2^2)(\rho^0\gamma^2 v + (\rho^0)'gv) + \gamma^2 \frac{d}{dx_3}\rho^0\frac{d}{dx_3}v = 0. \tag{12.1.16}$$

Replacing x_3 by x ($x \in \mathbb{R}$) and dividing by $\gamma^2 k^2$, with $k^2 = k_1^2 + k_2^2$, we obtain

$$\left[-\frac{1}{k^2}\frac{d}{dx}\rho^0\frac{d}{dx} + \rho^0 + (\rho^0)'\frac{g}{\gamma^2}\right]v = 0. \tag{12.1.17}$$

If we look for solutions $\hat{p}$, $\hat{u}_1$, $\hat{u}_2$, and $\hat{\rho}$ in the form (12.1.15), we obtain these solutions by returning to the initial equations (12.1.4)–(12.1.7):

$$\hat{p}(x_1, x_2, x_3, t; k_1, k_2) = p(x_3) \exp(\gamma t + ik_1x_1 + ik_2x_2),$$

$$\hat{u}_1(x_1, x_2, x_3, t; k_1, k_2) = -\frac{k_1}{\rho^0(x_3)\gamma} p(x_3) \exp(\gamma t + ik_1x_1 + ik_2x_2),$$

$$\hat{u}_2(x_1, x_2, x_3, t; k_1, k_2) = -\frac{k_2}{\rho^0(x_3)\gamma} p(x_3) \exp(\gamma t + ik_1x_1 + ik_2x_2),$$

$$\hat{\rho}(x_1, x_2, x_3, t; k_1, k_2) = -\frac{(\rho^0)'(x_3)}{\gamma} p(x_3) \exp(\gamma t + ik_1x_1 + ik_2x_2),$$

where

$$p(x_3) = -\frac{1}{k_1^2 + k_2^2}\rho^0(x_3)\gamma v'(x_3).$$

Hence we are left with the problem of finding a triple (k, γ, v) such that (12.1.17) is satisfied. There are two relevant parameters,

$$h = \frac{1}{k}$$

and

$$\delta = \frac{g}{\gamma^2}.$$

The existence of a triple with $\gamma > 0$ is a sign of an instability as $t \to +\infty$.

12.2 The mathematical question

So, in view of the discussion in Section 12.1, the mathematical problem is to analyze the kernel of the following operator as a function of $\delta \in \mathbb{R}$:

$$P(h, \delta) := -h^2 \frac{d}{dx} \rho(x) \frac{d}{dx} + \rho(x) + \delta \rho'(x), \qquad (12.2.1)$$

where $\rho = \rho^0$ is obtained from the stationary solution of the initial system (12.1.1). Although standard spectral theory does not apply here, we shall see that the same techniques apply. Here $h > 0$, and $\rho(x)$ is a C^1 function with the following properties:

$$\lim_{x \to -\infty} \rho(x) = \rho_1 > 0, \qquad \lim_{x \to +\infty} \rho(x) = \rho_2 > 0, \qquad (12.2.2)$$

and

$$\rho(x) > 0, \ \forall x \in \mathbb{R}. \qquad (12.2.3)$$

We also assume that ρ' satisfies

$$\lim_{|x| \to +\infty} \rho'(x) = 0, \qquad (12.2.4)$$

and that

$$\rho_1 \neq \rho_2. \qquad (12.2.5)$$

Note that this implies in particular that $\rho'(x)$ cannot be identically 0. Also, the question that we are interested in may be semiclassical (that is, $h \to 0$) or it may correspond to the asymptotics of $h \to +\infty$. Finally, note that the most physical case is the case where ρ' is negative and, in particular,

$$\rho_2 < \rho_1. \qquad (12.2.6)$$

12.3 Generalized spectrum

We first observe that there is no problem, using what we have done for the Schrödinger operator in Section 9.4, in defining the self-adjoint extension of $P(h, \delta)$ (we keep the same notation) with domain $H^2(\mathbb{R})$, and it can be immediately observed that $P(h, 0)$ is injective (and positive).

Definition 12.1 The "generalized spectrum" of the family $P(h, \delta)$ is the set of δ's such that $P(h, \delta)$ is not injective.

Remark 12.2 A standard analysis of the solutions of ordinary differential equations at ∞ (see, for example, [Sib]) shows that for any δ, the dimension of $N(P(h, \delta))$ is either zero or one.

The first result is the following.

Proposition 12.3 *Under the above assumptions and assuming that ρ' is not identically 0, the set of δ's such that $N(P(h, \delta))$ is not zero can be described by two sequences (possibly empty or finite) δ_n^+ and δ_n^- such that*

$$0 < \delta_n^+ < \delta_{n+1}^+,$$
$$\lim_{n \to +\infty} \delta_n^+ = +\infty \tag{12.3.1}$$

(the second condition being applied only in the infinite case), and

$$0 < -\delta_n^- < -\delta_{n+1}^-,$$
$$\lim_{n \to +\infty} \delta_n^- = -\infty \tag{12.3.2}$$

(the fourth condition being applied only in the infinite case).

Proof We observe that

$$N(P(h, \delta)) \neq \{0\} \text{ if and only if } N\left(K(h) + \frac{1}{\delta}\right) \neq \{0\}, \tag{12.3.3}$$

with

$$K(h) = P(h, 0)^{-1/2} \rho'(x) P(h, 0)^{-1/2}.$$

Equivalently, $N(P(h, \delta)) \neq \{0\}$ if and only if $-1/\delta$ is an eigenvalue of $K(h)$. So, the problem is immediately reduced to the spectral analysis of a self-adjoint compact operator $K(h)$, as presented in Section 6.4. $\square$

Remark 12.4 In fact, some of the sequences may be empty. Here are some particular cases.

- If $\rho'(x)$ is identically 0, then the spectrum of $K(h)$ is zero, and the set of δ such that $P(h, \delta)$ is not injective is empty.
- If $\rho'(x) > 0$, then we have only the negative sequence δ_n^-.
- If ρ' vanishes in an open set, then 0 is in the spectrum of $K(h)$. However, we are interested only in the nonzero eigenvalues of $K(h)$.

We also note that a particular role is played by the quantities $\inf_{||u||=1}\langle K(h)u, u\rangle$ and $\sup_{||u||=1}\langle K(h)u, u\rangle$. By Theorem 6.18, we know that they belong to the spectrum of $K(h)$. We observe that if δ_1^+ exists, then it is given by

$$\frac{1}{\delta_1^+} = - \inf_{||u||=1} \langle K(h)u, u\rangle. \tag{12.3.4}$$

Similarly, if δ_1^- exists, then it is given by

$$\frac{1}{\delta_1^-} = - \sup_{||u||=1} \langle K(h)u, u\rangle. \tag{12.3.5}$$

Up to this point, we have been doing abstract analysis, and $h > 0$ has been considered as a fixed parameter.

The next observation (see Theorem 6.18) is that

$$\sup_{||u||=1} |\langle K(h)u, u\rangle| \leq \Lambda := \frac{\sup_x |\rho'(x)|}{\inf_x \rho(x)}.$$

As a consequence, we see that in the interval $I :=]-1/\Lambda, +1/\Lambda[$, which is independent of h, we have

$$N(P(h, \delta)) = \{0\}, \ \forall \delta \in I. \tag{12.3.6}$$

A different estimate can be obtained rather easily. If u is a solution for some $\delta \neq 0$, then we obtain, by taking the scalar product in L^2 with u,

$$\int_{-\infty}^{+\infty} \rho(h^2u'(x)^2 + u(x)^2)\, dx = -\delta \int_{-\infty}^{+\infty} \rho'(x)u(x)^2\, dx$$
$$= 2\delta \int_{-\infty}^{+\infty} \rho u(x)u'(x)\, dx. \tag{12.3.7}$$

Using the Cauchy–Schwarz inequality, we obtain

$$\int_{-\infty}^{+\infty} \rho(x) \left(1 - \frac{|\delta|}{h}\right) (h^2 u'(x)^2 + u(x)^2)\, dx \leq 0. \qquad (12.3.8)$$

So, we have shown the following proposition.

Proposition 12.5

$$N(P(h, \delta)) = \{0\}, \ \forall \delta \in \,] - h, h[\,. \qquad (12.3.9)$$

We observe that this estimate is not interesting in the semiclassical regime (except when $\inf_x \rho(x) = 0$).

Exercise 12.1 Assuming that $\rho' > 0$, show the same results by using a proof similar to the proof of the uncertainty principle given in Chapter 1 (see around (1.3.14)). Use the change of function $u = \rho^{1/2} v$ and the differential operators $X_1 = \rho^{1/2}(d/dx)\rho^{1/2}$ and $X_2 = \rho$.

12.4 The initial toy model

In the case where $\rho(x) = \rho_1$ for $x < 0$ and $\rho(x) = \rho_2$ for $x > 0$ with $\rho_1 \neq \rho_2$, the generalized eigenvalue is unique and is given by

$$\delta^c(h) = h(\rho_1 + \rho_2)/(\rho_1 - \rho_2). \qquad (12.4.1)$$

This model was investigated by Rayleigh in 1893. There are at least two points of view from which this problem can be understood.

The first point of view is a good exercise in distribution theory and consists in using the general description of solutions in $L^2(\mathbb{R}^{\pm})$ for $\pm x > 0$. For $x > 0$, the solution should be $C_+ \exp -(1/h)x$, and for $x < 0$ the solution should be $C_- \exp(1/h)x$. A possible generalized eigenfunction, for some δ, is found by imposing the condition that the two functions can be matched in the form of a function u in $H^1(\mathbb{R})$ satisfying (in the sense of distributions) $P(h, \delta)u = 0$. Here, $\rho'(x)$ should be understood as the following operator (well defined on $H^1(\mathbb{R})$): $u \mapsto (\rho_2 - \rho_1)u(0)\mu_0$, where μ_0 is the Dirac measure at the origin. So, we take $u(x) = \exp -(1/h)|x|$ and look for δ's such that $P(h, \delta)u = 0$. We now note that

$$\left(-h^2 \frac{d}{dx}\rho(x)\frac{d}{dx} + \rho(x)\right) u = h(\rho_1 + \rho_2)\mu_0.$$

This gives a method for computation.

The second point of view is to observe that the associated compact operator K is actually of rank one. On the one hand, the range of K can be described as the vector space generated by $(-h^2(d/dx)\rho(d/dx) + \rho)^{-1/2}\mu_0$. On the other hand, we know that it is also related to the kernel of $P(h, \delta)$ through the map

$$v \mapsto Sv = P(h, 0)^{-1/2}v.$$

Eigenvectors u are easier to compute by the first approach, and the associated v generating the range of K is then determined by the map S^{-1}.

12.5 Toward a more realistic model

We now suppose that ρ is equal to ρ_1 and ρ_2 in the neighborhood of $\pm\infty$. We analyze an asymptotic regime corresponding to a smooth transition between ρ_1 and ρ_2 over an interval of size ϵ. The initial problem is to find δ such that the operator

$$P_\epsilon(\delta) = -\frac{d}{dx}\rho_\epsilon\frac{d}{dx} + \rho_\epsilon(x) + \delta\rho_\epsilon'(x), \qquad (12.5.1)$$

where $\rho_\epsilon(x) = \rho(x/\epsilon)$, has a nonzero kernel.

If we assume that ρ is constant outside $[-1, +1]$, the analysis of the L^2 solutions outside this interval is easy. We prefer to look at an analysis for a scaled operator

$$H(\epsilon, \delta) = -\frac{d}{dx}\rho\frac{d}{dx} + \epsilon^2\rho(x) + \delta\epsilon\rho'(x). \qquad (12.5.2)$$

We would like to analyze the problem in the limit $\epsilon \to 0$. Using an explicit description of the solutions outside $[-1, +1]$ (the solution is equal to $A_+ \exp -\epsilon x$ in $]1, +\infty[$ and consequently satisfies $u' = -\epsilon u$ in this interval, and, similarly, the solution is equal to $A_- \exp \epsilon x$ in $]-\infty, -1[$ and consequently satisfies $u' = \epsilon u$ in $]-\infty, -1[$), the problem is reduced to an analysis in $]-1, +1[$ of the following problem:

$$H(\epsilon, \delta)u = 0,$$
$$u'(+1) = -\epsilon u(1),$$
$$u'(-1) = +\epsilon u(-1).$$

We note first, assuming that δ has a formal expansion in powers of ϵ, that we can perform a formal expansion of the system in powers of ϵ that can be solved by recursion.

On the other hand, this is also a consequence of some rather standard perturbation theory. In order to treat this problem, it is useful to eliminate the dependence of the boundary problem on ϵ. But this is easily done by writing $u = e^{-\sqrt{2\epsilon}\langle x\rangle}v$ and by considering

$$\hat{H}(\epsilon,\delta) := e^{\sqrt{2\epsilon}\langle x\rangle}H(\epsilon,\delta)e^{-\sqrt{2\epsilon}\langle x\rangle},$$

and the Neumann condition at ± 1. Here, $\langle x\rangle$ is the Japanese bracket $\langle x\rangle = \sqrt{1+x^2}$. We then have to analyze the spectrum near 0 of the operator corresponding to the Neumann realization in $]-1,+1[$ of

$$\tilde{H}(\epsilon,\gamma) := -\left(\frac{d}{dx} - \epsilon\sqrt{2}\frac{x}{\langle x\rangle}\right)\rho\left(\frac{d}{dx} - \epsilon\sqrt{2}\frac{x}{\langle x\rangle}\right) + \epsilon^2\rho + \gamma\rho',$$

as a function of ϵ and γ, where $\gamma = \delta\epsilon$.

For $\epsilon = \gamma = 0$, $\tilde{H}(0,0)$ has 0 as an isolated eigenvalue of multiplicity 1, with $u_0 = 1/\sqrt{2}$ as the corresponding eigenfunction. So, by regular perturbation theory, the eigenvalue (near 0) $F(\epsilon,\gamma)$ depends analytically on ϵ and γ. In addition, the so-called Feynman–Hellmann formula states that

$$\left(\frac{\partial F}{\partial\gamma}\right)(0,0) = \left\langle\left(\frac{\partial\tilde{H}}{\partial\gamma}\right)(0,0)u_0,u_0\right\rangle_{L^2(]-1,+1[)} \quad ;$$

that is,

$$\left(\frac{\partial F}{\partial\gamma}\right)(0,0) = \frac{1}{2}\int_{-1}^{+1}\rho'(x)\,dx = \frac{1}{2}(\rho_2 - \rho_1) \neq 0.$$

This permits us to apply the implicit function theorem, and we obtain $\gamma(\epsilon)$ such that $F(\gamma(\epsilon),\epsilon) = 0$. We observe that $\gamma(0) = 0$. We then recover

$$\delta = \delta^c + \sum_{j\geq 1}c_j\epsilon^j, \qquad (12.5.3)$$

by dividing by ϵ.

In order to obtain the main term in (12.5.3), we compute $(\partial F/\partial\epsilon)(0,0)$. We obtain

$$\left(\frac{\partial F}{\partial\epsilon}\right)(0,0) = \frac{1}{2}\left\langle\left(\frac{\partial\tilde{H}}{\partial\epsilon}\right)_{\epsilon=0}u_0,u_0\right\rangle = \frac{1}{2}(\rho_1 + \rho_2).$$

This gives

$$\delta^c = \frac{\rho_1 + \rho_2}{\rho_1 - \rho_2}. \qquad (12.5.4)$$

12.6 Notes

The material of this chapter is extracted from a joint paper with O. Lafitte [Hel-Laf], which was a continuation of the papers [ChLa], [Laf], and [CCLaRa]. A much more general case was considered in those papers. For Section 12.3, readers can consult Sibuya's book [Sib] for the basic properties of ordinary differential equations. In the case of Section 12.5, we refer readers to [Ka] for details of regular perturbation theory.

13
Non-self-adjoint operators and pseudospectra

13.1 Preliminaries

When an operator is not self-adjoint, the spectrum is not the right object to consider, in the sense that it becomes very unstable with respect to perturbations. It has been realized in recent years that a family of sets (parametrized by $\epsilon > 0$) in $\mathbb{C}$ called the ϵ-pseudospectrum (or pseudospectra) is the right object for obtaining stability in this case. Similarly, a knowledge of the spectrum is not sufficient for analyzing the behavior of the associated semigroup in the non-self-adjoint case. A typical question that is easy in the self-adjoint case and difficult in the non-self-adjoint case is the following. If we consider a semibounded self-adjoint operator A on a Hilbert space $\mathcal{H}$, we have seen in Section 8.6 (last example) that by the functional calculus, we can define for any $t > 0$ the operator $\exp -tA$ (i.e., $f(A)$ with $f(\lambda) = e^{-t\lambda}$). We immediately obtain

$$\|e^{-tA}\|_{\mathcal{L}(\mathcal{H})} = e^{-t\,\inf\sigma(A)}, \tag{13.1.1}$$

and we recall that

$$\|(A - z)^{-1}\| = \frac{1}{d(z, \sigma(A))}. \tag{13.1.2}$$

In the case of a non-self-adjoint operator, however, we may lose the equality, and have only

$$\|(A - z)^{-1}\| \geq \frac{1}{d(z, \sigma(A))}. \tag{13.1.3}$$

For example, in the case where $\mathcal{H} = \mathbb{C}^2$ and

$$A = \begin{pmatrix} 0 & 1 \\ 0 & 0 \end{pmatrix},$$

164

we have $\sigma(A) = \{0\}$, but

$$e^{-tA} = \begin{pmatrix} 1 & -t \\ 0 & 1 \end{pmatrix}$$

does not satisfy (13.1.1), and (13.1.2) does not hold. In fact, $||(A - z)^{-1}||$ behaves like $1/|z|^2$ as $z \to 0$. Hence, we need new concepts in the non-self-adjoint case, and we shall explain some of them in this chapter.

13.2 Main definitions and properties

Definition 13.1 If A is a closed operator with a dense domain $D(A)$ in a Hilbert space $\mathcal{H}$ and if $\epsilon > 0$, then the ϵ-pseudospectrum $\sigma_\epsilon(A)$ of A is defined by

$$\sigma_\epsilon(A) := \left\{ z \in \mathbb{C} \mid ||(zI - A)^{-1}|| > \frac{1}{\epsilon} \right\},$$

where we adopt the convention that $||(zI - A)^{-1}|| = +\infty$ if $z \in \sigma(A)$, $\sigma(A)$ denoting the spectrum of A.

We shall be interested in this notion mainly in the limit $\epsilon \to 0$. It is clear that we always have

$$\sigma(A) \subset \sigma_\epsilon(A). \tag{13.2.1}$$

When A is self-adjoint (or, more generally, normal), $\sigma_\epsilon(A)$ is given by

$$\sigma_\epsilon(A) = \{z \in \mathbb{C} \mid d(z, \sigma(A)) < \epsilon\},$$

by the spectral theorem (see Proposition 8.20). In the non-self-adjoint case, using (13.1.3), we keep only

$$\{z \in \mathbb{C} \mid d(z, \sigma(A)) < \epsilon\} \subset \sigma_\epsilon(A). \tag{13.2.2}$$

So, it is only in the case of non-self-adjoint operators that this concept can give additional information about the spectral properties of the operator.

Although formulated in a rather abstract way, the following theorem explains rather well what the ϵ-pseudospectrum corresponds to.

Theorem 13.2 *(Roch–Silbermann theorem.)*

$$\sigma_\epsilon(A) = \bigcup_{\{\delta A \in \mathcal{L}(\mathcal{H}) \text{ s.t. } ||\delta A||_{\mathcal{L}(\mathcal{H})} < \epsilon\}} \sigma(A + \delta A).$$

In other words, z is in the ϵ-pseudospectrum of A if z is in the spectrum of some perturbation $A + \delta A$ of A with $||\delta A|| < \epsilon$. This is a natural notion. The models that we analyze are only approximations of real life, and the numerical analysis of these models proceeds through an analysis of explicitly computable approximate problems.

Proof Let us first show the easy part of this characterization of the ϵ-pseudospectrum. If $||(z-A)^{-1}|| \leq 1/\epsilon$, it is clear that for any δA such that $||\delta A|| < \epsilon$, $(A + \delta A - z)$ is invertible. Its inverse is obtained by observing first that

$$(z - A)^{-1}(z - A - \delta A) = I - (z - A)^{-1}\delta A.$$

But the left-hand side is invertible because

$$||(z - A)^{-1}\delta A|| \leq ||(z - A)^{-1}|| \, ||\delta A|| < 1.$$

Consequently, the left inverse is given by

$$(z - A - \delta A)^{-1} = \left(\sum_j ((z - A)^{-1} \delta A)^j \right) (z - A)^{-1}.$$

We can proceed similarly for the right inverse, and it can be seen immediately that the left inverse equals the right inverse, which is a general fact for operators that are left- and right-invertible.

The converse is not very difficult. If $z \in \sigma(A)$, then $\delta A = 0$ is a convenient choice. If $z \notin \sigma(A)$ and $||(z-A)^{-1}|| > 1/\epsilon$, then there exists, by the definition of the norm on $\mathcal{L}(\mathcal{H})$, some $u \in \mathcal{H}$ such that $||u|| = 1$ and

$$||(z - A)^{-1}u|| = \mu > \frac{1}{\epsilon}.$$

Let $v = (z - A)^{-1}u$. Let δA be the bounded operator defined by

$$(\delta A)x = \mu^{-2}u \langle x, v \rangle, \quad \forall x \in \mathcal{H}.$$

It is clear that v belongs, by construction, to $D(A)$. An elementary computation shows that v is an eigenfunction of $A + \delta A$ associated with z and that $||\delta A|| = 1/mu < \epsilon$. Hence we have found a perturbation $A + \delta A$ of A such that $z \in \sigma(A + \delta A)$ and $||\delta A|| < \epsilon$. $\qquad\square$

Remark 13.3 Another way to present the ϵ-pseudospectrum is to say that $z \in \sigma_\epsilon(A)$ if and only if either $z \in \sigma(A)$ or there exists an ϵ-pseudoeigenfunction, that is, a $u \in D(A)$ such that $||u|| = 1$ and $||(z - A)u|| < \epsilon$. This is very often a way to obtain estimates of pseudospectra.

Theorem 13.4 (ϵ-*pseudospectrum of the adjoint.*) *For any closed densely defined operator A and any* $\epsilon > 0$, *we have*

$$\sigma_\epsilon(A^*) = \overline{\sigma_\epsilon(A)},$$

where, for a subset Σ *in* $\mathbb{C}$, *we denote by* $\overline{\Sigma}$ *the set*

$$\overline{\Sigma} = \{z \in \mathbb{C} \mid \bar{z} \in \Sigma\}.$$

Proof This is immediate using the fact that if $z \notin \sigma(A)$, then

$$\|(z - A)^{-1}\| = \|(\bar{z} - A^*)^{-1}\|. \qquad \square$$

Other elementary properties

Theorem 13.5 *For* $\epsilon > 0$, *an operator A as defined above, and* $E \in \mathcal{L}(\mathcal{H})$ *such that* $\|E\| < \epsilon$, *we have*

$$\sigma_{\epsilon - \|E\|}(A) \subset \sigma_\epsilon(A + E) \subset \sigma_{\epsilon + \|E\|}(A).$$

Proof Suppose that $z \in \sigma_{\epsilon - \|E\|}(A)$. Then there exists F with $\|F\| < \epsilon - \|E\|$ such that $z \in \sigma(A+F) = \sigma((A+E)+(F-E))$. If we observe that $\|F-E\| \leq \|F\| + \|E\| < \epsilon$, we obtain the result that $z \in \sigma_\epsilon(A + E)$.

The second inclusion follows by exchanging the roles of A and $A + E$. $\square$

Proposition 13.6 *Suppose A is an* $n \times n$ *matrix with n distinct eigenvalues* $\sigma(A) = \{\lambda_j\}$. *We have, as* $\epsilon \to 0$,

$$\sigma_\epsilon(A) \subset \cup_j \left(D(\lambda_j, \epsilon\kappa(\lambda_j)) + \mathcal{O}(\epsilon^2) \right),$$

where $\kappa(\lambda_j) \geq 1$ *is defined below in* (13.2.3).

Proof According to Theorem 13.2, we have to analyze the spectrum of $A(t) = A + tE$, where E is a matrix with a norm less than or equal to 1. We start from a biorthogonal basis:

$$A v_j = \lambda_j v_j, \; A^* u_j = \bar{\lambda}_j u_j.$$

We have $\langle u_j, v_k \rangle = 0$ for $j \neq k$, and for $j = k$ this implies $\langle u_j, v_j \rangle \neq 0$. Under the assumption of simplicity, the eigenvalues $\lambda_j(t)$ are a C^∞ function of t for $|t| \leq t_0$, and we can choose a basis of eigenfunctions $v_j(t)$ ($j = 1, \ldots, n$)

such that $v_j(0) = v_j$ and such that $t \mapsto v_j(t)$ belongs to $C^\infty(]-t_0, t_0[)$. We can do the same for A^*.

Writing

$$A(t)v_j(t) = \lambda_j(t)\, v_j(t),$$

differentiating with respect to t, and taking $t = 0$, we obtain

$$\lambda'_j(0) = \frac{\langle Ev_j, u_j \rangle}{\langle v_j, u_j \rangle}.$$

Hence we define

$$\kappa(\lambda_j) := \frac{\|u_j\|\,\|v_j\|}{|\langle u_j, v_j \rangle|}, \tag{13.2.3}$$

and the Cauchy–Schwarz inequality implies that

$$\kappa(\lambda_j) \geq 1.$$

$\kappa(\lambda_j)$ is called the instability index. Note that in the self-adjoint case $\kappa(\lambda_j) = 1$. We can in fact take $u_j = v_j$.

We have not been very careful about controlling the uniformity with respect to E here. In the case considered here, this can be done using the fact that the closed unit ball in $\mathcal{L}(\mathbb{C}^N)$ is compact. $\qquad\square$

If we normalize by $\langle u_j, v_j \rangle = 1$, then $u \mapsto P_j u = v_j \langle u, u_j \rangle$ is a projector on the eigenspace associated with λ_j along (i.e., parallel to) $\mathrm{span}\{v_k\}_{k \neq j}$, and its norm is $\kappa(\lambda_j)$. For the extension to the infinite-dimensional case, it is useful to observe (using the diagonalization of A) that

$$P_j = \frac{1}{2\pi i} \int_{\Gamma_j} (z - A)^{-1}\, dz,$$

where Γ_j is a path that turns once in the region of positive sign such that the bounded component of $\mathbb{C} \setminus \mathrm{Im}\,\gamma$ intersected with $\sigma(A)$ contains only λ_j. This is called the spectral projector attached to the spectrum surrounded by Γ_j.

Remark 13.7 It can also be shown that

$$\cup_j \left(D(\lambda_j, \epsilon\kappa(\lambda_j)) + \mathcal{O}(\epsilon^2) \right) \subset \sigma_\epsilon(A).$$

For this, in the spirit of the proof of Proposition 13.6, we have to find, for each j, some operator E_j such that $\lambda'_j(0) = \kappa(\lambda_j)$. But it is enough to take $E_j = (1/\kappa(\lambda_j))\, P_j^*$.

Resolvent and subharmonic functions We recall that a real-valued continuous function on a domain $G \subset \mathbb{C}$ is subharmonic if, for any closed disk $D(z, r) \subset G$ with center z and radius r, we have

$$\varphi(z) \le \frac{1}{2\pi} \int_0^{2\pi} \varphi(z + re^{i\theta}) \, d\theta.$$

Another characterization in the C^2 case is that $\Delta \varphi \ge 0$. A third characterization is that ϕ is subharmonic if, for any $r > 0$, and any harmonic function h in $D(z, r)$ such that $\phi \le h$ on $S(z, r) = \partial D(z, r)$, $\phi \le h$ in $D(z, r)$.

From the definition given after (7.2.10), φ is subharmonic if and only if $-\varphi$ is superharmonic. We recall a general result.

Proposition 13.8 *If A is a closed operator, then the function $\psi(z) = \|(A - z)^{-1}\|$ is a subharmonic function on $\mathbb{C} \setminus \sigma(A)$.*

Proof For any $u, v \in \mathcal{H}$, we observe that $\langle (A - z)^{-1} u, v \rangle$ is a holomorphic function in $\mathbb{C} \setminus \sigma(A)$. This implies that $z \mapsto \operatorname{Re} \langle (A - z)^{-1} u, v \rangle$ is a harmonic function (we identify $z = x + iy$ in $\mathbb{C}$ with the point (x, y) in $\mathbb{R}^2$). Taking the supremum of this family of harmonic functions over (u, v) with $\|u\| \le 1$ and $\|v\| \le 1$ gives a subharmonic function (as can be immediately deduced from the third characterization). $\square$

Corollary 13.9 *Every bounded connected component of $\sigma_\epsilon(A)$ contains an element of $\sigma(A)$.*

If this were not the case, $\psi(z)$ would be subharmonic in such a component and we would have a contradiction with the definition of subharmonicity.

Remark 13.10 It can be shown similarly that $\log \psi(z)$ is subharmonic.

13.3 Accretive operators

Here, we collect together some material on accretive operators.

Definition 13.11 Let A be an unbounded operator in $\mathcal{H}$ with domain $D(A)$. We say that A is accretive if

$$\operatorname{Re} \langle Ax, x \rangle_{\mathcal{H}} \ge 0, \ \forall x \in D(A). \tag{13.3.1}$$

Proposition 13.12 *Let A be an accretive operator with a domain $D(A)$ that is dense in $\mathcal{H}$. Then A is closable, and its closed extension $\overline{A}$ is accretive.*

Proof Let x_n be a sequence in $D(A)$ such that $x_n \to 0$ and $Ax_n \to \ell$. We have to show that ℓ is 0. We proceed by contradiction. If $\ell \neq 0$, let $\eta \in D(A)$ such that $\langle \eta, \ell \rangle \neq 0$ (here we use the fact that $D(A)$ is dense in $\mathcal{H}$). To determine $c \in \mathbb{C}$, we consider the sequence $\mathrm{Re} \langle A(\eta + cx_n), \eta + cx_n \rangle$. Taking the limit as $n \to +\infty$, we obtain

$$\lim_{n \to +\infty} \mathrm{Re} \langle A(\eta + cx_n), \eta + cx_n \rangle = \mathrm{Re} \langle A\eta, \eta \rangle + \mathrm{Re} (c\langle \eta, \ell \rangle).$$

It can be seen immediately that we can choose c such that $\lim_{n \to +\infty} \mathrm{Re} \langle A(\eta + cx_n), \eta + cx_n \rangle < 0$, in contradiction with the accretivity of A. $\qquad\square$

Once we have this proposition, it is natural to ask about the uniqueness of this accretive extension. This leads to the following definition.

Definition 13.13 An accretive operator A is maximally accretive if an accretive extension $\tilde{A}$ with strict inclusion of $D(A)$ in $D(\tilde{A})$ does not exist.

The following criterion, which extends the standard criterion for essential self-adjointness (see Theorem 9.8), is the most suitable.

Theorem 13.14 *For an accretive operator A, the following conditions are equivalent:*

1. *$\overline{A}$ is maximally accretive.*
2. *There exists $\lambda_0 > 0$ such that $A^* + \lambda_0 I$ is injective.*
3. *There exists $\lambda_1 > 0$ such that the range of $A + \lambda_1 I$ is dense in $\mathcal{H}$.*

The proof is left to the reader, bearing in mind the discussion around (9.2.2), which can be replaced by

$$\mathrm{Re} \langle (A + b)u, u \rangle \geq b \, ||u||^2, \forall u \in D(A) \text{ and } b > 0.$$

Remark 13.15 Of course, a nonnegative essentially self-adjoint operator is maximally accretive, but it is not so easy to extend what we know in the self-adjoint case to this more general situation because we can no longer use the functional calculus based on the spectral theorem.

13.4 Introduction to semigroups, and the Hille–Yosida theorem

We do not pretend to present a complete course on this theory, which is the subject of many books, but will present only some useful results here. Most of the time, we shall give sketches of the proofs. Let us start with some simple natural definitions. If $\mathcal{H}$ is a Hilbert space, a one-parameter semigroup of operators on $\mathcal{H}$ is a family of operators indexed on the nonnegative real numbers $\{T(t)\}_{t \in [0, +\infty[}$ such that

$$T(0) = I, \; T(s + t) = T(s) \circ T(t), \; \forall t, s \geq 0. \tag{13.4.1}$$

Definition 13.16 A semigroup is said to be strongly continuous if, for any $x \in \mathcal{H}$, the mapping $t \mapsto T(t)x$ is continuous from $[0, +\infty[$ into $\mathcal{H}$.

Recall that by the Banach–Steinhaus theorem, $\sup_J \|T(t)\| =: m(J)$ is bounded for every compact interval $J \subset [0, +\infty[$. Using the semigroup property, the next proposition follows easily.

Proposition 13.17 *If $T(t)$ is a strongly continuous semigroup, then there exist $M \geq 1$ and $\omega_0 \in \mathbb{R}$ such that $T(t)$ has the property*

$$P(M, \omega_0): \quad \|T(t)\| \leq M e^{\omega_0 t}, \; t \geq 0. \tag{13.4.2}$$

In fact, we have this property for $0 \leq t < 1$; for larger values of t, we can write $t = [t] + r$, $[t] \in \mathbb{N}$, $0 \leq r < 1$, and $T(t) = T(1)^{[t]} T(r)$.

The infinitesimal generator of a one-parameter semigroup T is an unbounded operator A, where $D(A)$ is the set of the $x \in \mathcal{H}$ such that $h^{-1}(T(h)x - x)$ has a limit as h approaches 0 from the right. The value of Ax is the value of the above limit. In other words, Ax is the right-derivative at 0 of the function $t \mapsto T(t)x$.

Example 13.18 The simplest interesting example is when $A = \Delta$, $D(A) = H^2$, and $\mathcal{H} = L^2(\mathbb{R}^n)$, and $T(t)$ is defined, for $f \in L^2(\mathbb{R}^n)$, by

$$T(t)f = (4\pi t)^{-n/2} \int_{\mathbb{R}^n} e^{-(|x-y|^2)/t} f(y) \, dy.$$

But we can use the functional calculus for self-adjoint operators in this case.

Basics of semigroup theory

Theorem 13.19 *(Existence and uniqueness.) Suppose that A generates a strongly continuous semigroup $T(t)$ on $\mathcal{H}$. Then, for any $u_0 \in D(A)$, $u(t) := T(t)u_0$ satisfies:*

1. $u \in C^0([0, +\infty[, D(A)) \cap C^1([0, +\infty[, \mathcal{H})$.
2. $u'(t) = Au(t)$, $u(0) = u_0$.
3. $Au(t) = T(t)Au_0$.

Moreover, the solution of the second equation is unique.

Corollary 13.20 *Under the same assumptions, for any* $f \in C^0([0, T[, \mathcal{H})$ *and any* $u_0 \in D(A)$, *the unique solution of*

$$u'(t) - Au(t) = f(t), 0 < t < T, \ u(0) = u_0,$$

is given by

$$u(t) = T(t)u_0 + \int_0^t T(t - s)f(s)\,ds.$$

The Hille–Yosida theorem provides a necessary and sufficient condition for a closed linear operator A on a Hilbert space to be the infinitesimal generator of a strongly continuous one-parameter semigroup.

Theorem 13.21 *Let A be a linear operator defined on a linear subspace $D(A)$ of a Hilbert space $\mathcal{H}$, let ω be a real number, and let $M > 0$. Then A generates a strongly continuous semigroup T that satisfies $P(M, \omega)$ if and only if:*

1. *$D(A)$ is dense in $\mathcal{H}$.*
2. *Every real $\lambda > \omega$ belongs to the resolvent set of A and, for such λ and for all positive integers n,*

$$\|(\lambda I - A)^{-n}\| \leq \frac{M}{(\lambda - \omega)^n}. \tag{13.4.3}$$

In the general case, the Hille–Yosida theorem is mainly of theoretical importance, since the estimates of the powers of the resolvent operator that appear in the statement of the theorem cannot usually be checked in concrete examples. In the special case of a contraction semigroup ($M = 1$ and $\omega = 0$ in the above theorem), only the case $n = 1$ has to be checked, and the theorem becomes of some practical importance. An explicit statement of the Hille–Yosida theorem for contraction semigroups is contained in the following theorem (in the case $\omega = 0$).

Theorem 13.22 *Let A be a linear operator defined on a linear subspace $D(A)$ of a Hilbert space $\mathcal{H}$. Then A generates a continuous semigroup $T(t)$ satisfying $P(1, \omega)$ if and only if:*

1. *$D(A)$ is dense in $\mathcal{H}$.*
2. *Every real $\lambda > \omega$ belongs to the resolvent set of A and, for such λ,*

$$\|(\lambda I - A)^{-1}\| \leq \frac{1}{(\lambda - \omega)}. \tag{13.4.4}$$

Proof In one direction, the proof is based on the basic formula

$$(z - A)^{-1} = \int_0^\infty T(t)e^{-tz}\,dt, \tag{13.4.5}$$

which holds, when $P(M, \omega_0)$ is satisfied, for z in the open half-plane $\mathrm{Re}\,z > \omega_0$. We have in fact

$$
\begin{aligned}
A\left(\int_0^\infty T(t)e^{-tz}\,dt\right) &= \int_0^\infty AT(t)e^{-tz}\,dt \\
&= \int_0^\infty T'(t)e^{-tz}\,dt \\
&= -I + z\int_0^\infty T(t)e^{-tz}\,dt. \tag{13.4.6}
\end{aligned}
$$

More precisely, we first have to prove that for $u \in D(A)$,

$$A\left(\int_0^\infty T(t)e^{-tz}\,dt\right)u = -u + z\int_0^\infty T(t)u e^{-tz}\,dt. \tag{13.4.7}$$

Then, using the fact that A is closed and $D(A)$ is dense, we obtain the result that for $u \in \mathcal{H}$, $(\int_0^\infty T(t)e^{-tz}\,dt)u \in D(A)$, and (13.4.7) is true for any $u \in \mathcal{H}$. From this we obtain (13.4.5), which implies

$$\|(z - A)^{-1}\|_{\mathcal{L}(\mathcal{H})} \le M\int_0^\infty e^{(\omega - \mathrm{Re}\,z)t}\,dt,$$

and (13.4.4) follows.

More generally, we can use (differentiating with respect to z (13.4.5))

$$(z - A)^{-n} = \int_0^\infty (-t)^{n-1} T(t)e^{-tz}\,dt \tag{13.4.8}$$

in order to obtain (13.4.3).

Remark 13.23 What we have actually proved is that $P(M, \omega)$ implies that every $z \in \mathbb{C}$ such that $\mathrm{Re}\,z > \omega$ belongs to the resolvent set of A and satisfies, for any $n \ge 1$,

$$\|(zI - A)^{-n}\| \le \frac{M}{(\mathrm{Re}\,z - \omega)^n}. \tag{13.4.9}$$

It is then natural to ask about the converse, using the previous inequality (with $n = 1$) but for $z \in \mathbb{C}$. This kind of statement, which is different from the Hille–Yosida theorem, will be analyzed in the next section.

The converse part of the Hille–Yosida theorem   We shall not give a complete proof of the converse, but merely sketch the main step in a way that will permit us to see the difference between the two versions of the Hille–Yosida theorem. If we first reduce the problem to the case $\omega = 0$ by replacing A by $A - \omega$, we can use the following formula (which is not surprising, at least in the scalar case and in the case where $-A$ is a self-adjoint nonnegative operator):

$$T(t) = \text{s-}\lim_{n \to +\infty} \left(I - \frac{tA}{n} \right)^{-n}, \ \forall t > 0, \tag{13.4.10}$$

where "s- lim" means that the limit is in the strong sense:

$$T(t)x = \lim_{n \to +\infty} \left(I - \frac{tA}{n} \right)^{-n} x, \ \forall x \in \mathcal{H}, \ \forall t > 0.$$

For the first version (Theorem 13.21), we can then observe that by (13.4.3),

$$\left\| \left(I - \frac{tA}{n} \right)^{-n} \right\| \leq M.$$

For the second version (Theorem 13.22), we can use the fact that by (13.4.4),

$$\left\| \left(I - \frac{tA}{n} \right)^{-n} \right\| \leq \left\| \left(I - \frac{tA}{n} \right)^{-1} \right\|^{n} \leq 1. \qquad \square$$

Corollary 13.24 *If $-A$ is maximally accretive, then A is the generator of a contraction semigroup.*

Proof We first observe that for $\lambda > 0$, we have

$$\langle (-A + \lambda)u, \ u \rangle \geq \lambda \|u\|^2, \ \forall u \in D(A).$$

This implies that

$$\|(A - \lambda)u\| \geq \lambda \|u\|, \ \forall u \in D(A).$$

This inequality implies the injectivity of $(A - \lambda)$ for $\lambda > 0$, and since $(A - \lambda)$ is closed, it implies that the range of $(A - \lambda)$ is closed.

Finally, it remains to show that the range of $(A - \lambda)$ is dense, which is a consequence of the maximal accretivity of $-A$ (see Theorem 13.14). Hence, we obtain the result that $\lambda \notin \sigma(A)$ for $\lambda > 0$ and $\|(A + \lambda)^{-1}\| \leq 1/\lambda$. We can then apply the Hille–Yosida theorem. $\qquad\square$

We conclude this section with the analog for accretive operators of the Kato–Rellich theorem (Theorem 9.10) for self-adjoint operators.

Theorem 13.25 *Let A be a maximally accretive operator, and let B be an accretive operator with $D(A) \subset D(B)$. Suppose that there exist $a < 1$ and b such that*

$$\|Bx\| \leq a\,\|Ax\| + b\,\|x\|, \ \forall x \in D(A).$$

Then $(A + B)$ with domain $D(A)$ is maximally accretive.

13.5 The Gearhart–Prüss theorem

We next state the following theorem.

Theorem 13.26 *(Gearhart–Prüss theorem.) Let A be a closed operator with a dense domain $D(A)$ generating a strongly continuous semigroup $T(t)$, and let $\omega \in \mathbb{R}$. Assume that $\|(z - A)^{-1}\|$ is uniformly bounded in the half-plane $\operatorname{Re} z \geq \omega$. Then there exists a constant $M > 0$ such that $P(M, \omega)$ holds.*

Reformulation in terms of ϵ-spectra For a given accretive closed operator $\mathcal{A}$, we introduce, for any $\epsilon > 0$,

$$\widehat{\alpha}_\epsilon(\mathcal{A}) = \inf_{z \in \sigma_\epsilon(\mathcal{A})} \operatorname{Re} z. \tag{13.5.1}$$

It is obvious that

$$\widehat{\alpha}_\epsilon(\mathcal{A}) \leq \inf_{z \in \sigma(\mathcal{A})} \operatorname{Re} z. \tag{13.5.2}$$

We also define the following in $[-\infty, +\infty[$:

$$\widehat{\omega}_0(\mathcal{A}) = \lim_{t \to +\infty} \frac{1}{t} \log \|T(t)\|. \tag{13.5.3}$$

Note that later we shall meet cases where $\widehat{\omega}_0(\mathcal{A}) = -\infty$, corresponding to semigroups with a decay faster than exponential.

Using the previous theorem with $A = -\mathcal{A}$, we obtain the following statement.

Theorem 13.27 *(Gearhart–Prüss theorem reformulated.) Let $\mathcal{A}$ be a densely defined closed operator in a Hilbert space X such that $-\mathcal{A}$ generates a contraction semigroup. Then*

$$\lim_{\epsilon \to 0} \widehat{\alpha}_\epsilon(\mathcal{A}) = -\widehat{\omega}_0(\mathcal{A}). \tag{13.5.4}$$

This version is interesting because it reduces the question of the decay, which is basic in the question of stability, to an analysis of the ϵ-spectra of the operator for ϵ small.

Proof of Theorem 13.27 Let $\omega > \widehat{\omega}_0(\mathcal{A})$. Then there exists M such that $P(M, \omega)$ is satisfied. Using the proof of the Hille–Yosida theorem (see the basic formula (13.4.5)), we immediately obtain the result that

$$\sup_{\mathrm{Re}\, z > \omega} \|(z - A)^{-1}\| < +\infty.$$

This implies that for $\epsilon > 0$ small enough, $\{\mathrm{Re}\, z > \omega\} \cap \sigma_\epsilon(A) = \emptyset$. This gives $\omega > -\lim_{\epsilon \to 0} \widehat{\alpha}_\epsilon(\mathcal{A})$. Hence

$$\widehat{\omega}_0(\mathcal{A}) \geq -\lim_{\epsilon \to 0} \widehat{\alpha}_\epsilon(\mathcal{A}).$$

Conversely, if $\omega > -\lim_{\epsilon \to 0} \widehat{\alpha}_\epsilon(\mathcal{A})$, then the assumption of Theorem 13.26 is satisfied and we obtain $P(M, \omega)$. □

Technical remarks about the Gearhart–Prüss theorem Note that we can improve the conclusion of Theorem 13.26 a little. If the assumption of this theorem holds for some ω, then it is actually automatically true for some $\omega' < \omega$. We have the following lemma.

Lemma 13.28 *Let $\omega \in \mathbb{R}$. If, for some $\rho(\omega) > 0$, $\|(z - A)^{-1}\| \leq 1/\rho(\omega)$ for $\mathrm{Re}\, z > \omega$, then for every $\omega' \in]\omega - \rho(\omega), \omega]$ we have*

$$\|(z - A)^{-1}\| \leq \frac{1}{\rho(\omega) - (\omega - \omega')}, \quad \mathrm{Re}\, z > \omega'.$$

Proof This is simply a variant of the proof of the proposition expressing the fact that the set of the invertible operators in $\mathcal{L}(\mathcal{H})$ is open.

Let $\widetilde{z} \in \mathbb{C}$, $\mathrm{Re}\, \widetilde{z} > \omega$. Then, by the definition of ρ, $\|(\widetilde{z} - A)^{-1}\| \leq 1/\rho(\omega)$. For $z \in \mathbb{C}$ with $|z - \widetilde{z}| < \rho(\omega)$, we have

$$(z - A)(\widetilde{z} - A)^{-1} = 1 + (z - \widetilde{z})(\widetilde{z} - A)^{-1},$$

where

$$\|(z - \widetilde{z})(\widetilde{z} - A)^{-1}\| \le |z - \widetilde{z}|/\rho(\omega) < 1,$$

so $1 + (z - \widetilde{z})(\widetilde{z} - A)^{-1}$ is invertible and

$$\|(1 + (z - \widetilde{z})(\widetilde{z} - A)^{-1})^{-1}\| \le \frac{1}{1 - |z - \widetilde{z}|/\rho(\omega)}.$$

Hence z belongs to the resolvent set of A, and

$$(z-A)^{-1} = (\widetilde{z}-A)^{-1}(1+(z-\widetilde{z})(\widetilde{z}-A)^{-1})^{-1}, \quad \|(z-A)^{-1}\| \le \frac{1}{\rho(\omega) - |z - \widetilde{z}|}.$$

Now, if $z \in \mathbb{C}$ and $\operatorname{Re} z > \omega'$, we can find $\widetilde{z} \in \mathbb{C}$ with $\operatorname{Re} \widetilde{z} > \omega$, $|z - \widetilde{z}| < \omega - \omega'$, and the lemma follows. $\qquad\square$

Proposition 13.29 *Let*

$$\omega_0 = \inf\{\omega \in \mathbb{R} \text{ s.t. } \{\operatorname{Re} z > \omega\} \subset \rho(A) \text{ and } \sup_{\operatorname{Re} z > \omega} \|(z - A)^{-1}\| < \infty\}, \tag{13.5.5}$$

and for $\omega > \omega_0$, let $r(\omega)$ be defined by

$$\frac{1}{r(\omega)} = \sup_{\operatorname{Re} z > \omega} \|(z - A)^{-1}\|. \tag{13.5.6}$$

Then $r(\omega)$ is an increasing function of ω; $\omega - r(\omega) \ge \omega_0$; and for $\omega' \in [\omega - r(\omega), \omega]$, we have

$$r(\omega') \ge r(\omega) - (\omega - \omega'). \tag{13.5.7}$$

Moreover, if $\omega_0 > -\infty$, then $r(\omega) \to 0$ when $\omega \searrow \omega_0$.

Note that if $P(M, \omega_1)$ is satisfied, we already know that $\|(z - A)^{-1}\|$ is uniformly bounded in the half-plane $\operatorname{Re} z \ge \beta$ if $\beta > \omega_1$. If $\alpha \le \omega_1$, we see that $\|(z - A)^{-1}\|$ is uniformly bounded in the half-plane $\operatorname{Re} z \ge \alpha$, provided that

- we have this uniform boundedness on the line $\operatorname{Re} z = \alpha$;
- A has no spectrum in the half-plane $\operatorname{Re} z \ge \alpha$;
- $\psi(z) := \|(z - A)^{-1}\|$ does not grow too wildly in the strip $\alpha \le \operatorname{Re} z \le \beta$: $\|(z - A)^{-1}\| \le C \exp(C \exp(k|\operatorname{Im} z|))$, where $k < \pi/(\beta - \alpha)$ and $C > 0$.

We then also have

$$\sup_{\operatorname{Re} z \geq \alpha} \|(z - A)^{-1}\| = \sup_{\operatorname{Re} z = \alpha} \|(z - A)^{-1}\|. \tag{13.5.8}$$

This follows from the subharmonicity of $\ln \|(z - A)^{-1}\|$, Hadamard's theorem (or the Phragmén–Lindelöf theorem in exponential coordinates), and the maximum principle.

The main result, which is due to Helffer and Sjöstrand, is the following theorem.

Theorem 13.30 (*Quantitative Gearhart–Prüss theorem.*) *Under the assumptions of Theorem 13.26, let $m(t)$ be a continuous positive function such that*

$$\|T(t)\| \leq m(t). \tag{13.5.9}$$

Then, for all triples $t, a, \tilde{a} > 0$ such that $t = a + \tilde{a}$, we have

$$\|T(t)\|_{\mathcal{L}(\mathcal{H})} \leq \frac{e^{\omega t}}{r(\omega) \, \|1/m\|_{e^{-\omega \cdot} L^2(]0,a[)} \, \|1/m\|_{e^{-\omega \cdot} L^2(]0,\tilde{a}[)}}. \tag{13.5.10}$$

In (13.5.10), we have the natural norm in the exponentially weighted space $e^{-\omega \cdot} L^2(]0, a[)$, and similarly with $\tilde{a}$ instead of a: $\|f\|_{e^{-\omega \cdot} L^2(]0,a[)} = \|e^{\omega \cdot} f(\cdot)\|_{L^2(]0,a[)}$.

Theorem 13.30 implies Theorem 13.26 Here, we shall just give a short argument that allows us to find some M. More accurate estimates will be given in Section 13.7. It is enough to check $t \geq 2$. We also observe that (13.5.10) is true more generally for $t \geq a + \tilde{a}$. Hence we can take $a = \tilde{a} = 1$, and (13.5.10) reads, for $t \geq 2$,

$$\|T(t)\|_{\mathcal{L}(\mathcal{H})} \leq \frac{e^{\omega t}}{r(\omega) \, \|1/m\|^2_{e^{-\omega \cdot} L^2(]0,1[)}}. \tag{13.5.11}$$

We obtain $P(M, \omega)$, with

$$M = \max \left(\sup_{t \leq 2} e^{-\omega t} \|T(t)\|, \ \frac{1}{r(\omega) \, \|1/m\|^2_{e^{-\omega \cdot} L^2(]0,1[)}} \right).$$

13.6 Proof of the quantitative Gearhart–Prüss theorem

We shall use the inhomogeneous equation

$$(\partial_t - A)u = w \text{ on } \mathbb{R}. \tag{13.6.1}$$

We shall use the results of Theorem 13.19 freely. Let $C^0_+(\mathcal{H})$ denote the subspace of all $v \in C^0(\mathbb{R}; \mathcal{H})$ that vanish near $-\infty$. For $k \in \mathbb{N}$, we define $C^k_+(\mathcal{H})$ and $C^k_+(\mathcal{D}(A))$ similarly. For $w \in C^0_+(\mathcal{H})$, we define $Ew \in C^0_+(\mathcal{H})$ by

$$Ew(t) = \int_{-\infty}^{t} T(t - s)w(s)\,ds. \tag{13.6.2}$$

It is easy to see that E is continuous from $C^k_+(\mathcal{H})$ into $C^k_+(\mathcal{H})$ and from $C^k_+(\mathcal{D}(A))$ into $C^k_+(\mathcal{D}(A))$. If $w \in C^1_+(\mathcal{H}) \cap C^0_+(\mathcal{D}(A))$, then $u = Ew$ is the unique solution of (13.6.1) in the same space. More precisely, we have

$$(\partial_t - A)Ew = w, \quad E(\partial_t - A)u = u, \tag{13.6.3}$$

for all $u, w \in C^1_+(\mathcal{H}) \cap C^0_+(\mathcal{D}(A))$.

Now recall that by (13.4.2) we have $P(M, \omega_0)$ for some M, ω_0. If $\omega_1 > \omega_0$ and $w \in C^0_+(\mathcal{H}) \cap e^{\omega_1} L^2(\mathbb{R}; \mathcal{H})$, then Ew belongs to the same space and, using Schur's lemma (see Lemma 7.1), we obtain

$$\|Ew\|_{e^{\omega_1} L^2(\mathbb{R};\mathcal{H})} \leq \left(\int_0^\infty e^{-\omega_1 t} \|T(t)\|\,dt \right) \|w\|_{e^{\omega_1} L^2(\mathbb{R};\mathcal{H})}$$

$$\leq \frac{M}{\omega_1 - \omega_0} \|w\|_{e^{\omega_1} L^2(\mathbb{R};\mathcal{H})}.$$

Now we consider Laplace transforms.

If $u \in e^{\omega \cdot} S(\mathbb{R}; \mathcal{H})$, then the Laplace transform of u,

$$\widehat{u}(\tau) = (2\pi)^{-1/2} \int_{-\infty}^{+\infty} e^{-t\tau} u(t)\,dt,$$

is well defined in $S(\Gamma_\omega; \mathcal{H})$, where

$$\Gamma_\omega = \{\tau \in \mathbb{C};\ \mathrm{Re}\,\tau = \omega\}$$

(this is simply the Fourier transform of $e^{-\omega \cdot} u$ at $\mathrm{Im}\,\tau$), and we have Plancherel's formula (see (2.1.7))

$$\|\widehat{u}\|^2_{L^2(\Gamma_\omega)} = \|u\|^2_{e^{\omega \cdot} L^2}. \tag{13.6.4}$$

Now we make the assumptions of Theorem 13.30, define ω and $r(\omega)$ as there, and let M, ω_0 be as above. Let $w \in e^{\omega \cdot} S_+(\mathcal{D}(A))$, where $S_+(\mathcal{D}(A))$ is, by definition, the space of all $u \in S(\mathbb{R}; \mathcal{D}(A))$, vanishing near $-\infty$. Then $w \in e^{\omega_1} S_+(\mathcal{D}(A))$ for all $\omega_1 \geq \omega$. If $\omega_1 > \omega_0$, then $u := Ew$ belongs to $e^{\omega_1 \cdot} S_+(\mathcal{D}(A))$ and solves (13.6.1). Taking the Laplace transform of that equation, we obtain

$$(\tau - A)\widehat{u}(\tau) = \widehat{w}(\tau), \qquad (13.6.5)$$

for $\operatorname{Re} \tau > \omega_0$. Note here that $\widehat{w}(\tau)$ is continuous in the half-plane $\operatorname{Re} \tau \geq \omega$, it is holomorphic in $\operatorname{Re} \tau > \omega$, and $\widehat{w}_{|\Gamma_{\widetilde{\omega}}} \in \mathcal{S}(\Gamma_{\widetilde{\omega}})$ for every $\widetilde{\omega} \geq \omega$. We use the assumptions of the theorem to write

$$\widehat{u}(\tau) = (\tau - A)^{-1}\widehat{w}(\tau), \qquad (13.6.6)$$

and we see that $\widehat{u}(\tau)$ can be extended to the half-plane $\operatorname{Re} \tau \geq \omega$ with the same properties as $\widehat{w}(\tau)$. By the Laplace (Fourier) inversion formula applied to Γ_ω, we conclude that $u \in e^{\omega \cdot} S_+(\mathcal{D}(A))$. Moreover, since

$$\|\widehat{u}(\tau)\|_{\mathcal{H}} \leq \frac{1}{r(\omega)} \|\widehat{w}(\tau)\|_{\mathcal{H}}, \ \tau \in \Gamma_\omega,$$

we obtain from Plancherel's formula the result that

$$\|u\|_{e^{\omega \cdot} L^2} \leq \frac{1}{r(\omega)} \|w\|_{e^{\omega \cdot} L^2}. \qquad (13.6.7)$$

Using the density of $D(A)$ in $\mathcal{H}$ together with standard cutoff and regularization arguments, we see that (13.6.7) extends to the case where $w \in e^{\omega \cdot} L^2(\mathbb{R}; \mathcal{H}) \cap C_+^0(\mathcal{H})$, leading to the result that $u := Ew$ belongs to the same space and satisfies (13.6.7).

Returning to the Cauchy problem, we now consider $u(t) = T(t)v$, for $v \in D(A)$, solving

$$\begin{cases} (\partial_t - A)u = 0, \ t \geq 0, \\ u(0) = v. \end{cases}$$

Let χ be a continuous, decreasing, piecewise C^1 function on $\mathbb{R}$, equal to 1 on $]-\infty, 0]$ and vanishing near $+\infty$. Then

$$(\partial_t - A)(1 - \chi)u = -\chi'(t)u,$$

and, bearing in mind that $\|u(t)\| \leq m(t)\|v\|$ by assumption, we obtain

$$\|\chi' u\|_{e^{\omega \cdot} L^2}^2 = \int_0^{+\infty} |\chi'(t)|^2 \|u(t)\|^2 e^{-2\omega t} \, dt$$
$$\leq \|\chi' m\|_{e^{\omega \cdot} L^2}^2 \|v\|^2,$$

where we notice that $\chi' m$ is well defined on $\mathbb{R}$ since $\operatorname{supp} \chi' \subset [0, \infty[$.

Now $(1 - \chi)u$ and $\chi'u$ are well defined on $\mathbb{R}$, and so, by (13.6.7),

$$\|(1 - \chi)u\|_{e^{\omega \cdot} L^2} \leq r(\omega)^{-1} \|\chi'u\|_{e^{\omega \cdot} L^2} \leq r(\omega)^{-1} \|\chi'm\|_{e^{\omega \cdot} L^2} \|v\|. \quad (13.6.8)$$

In fact, in order to apply (13.6.7), we first approximate χ by a sequence χ_n of smooth functions satisfying the same condition and then take the limit. Hence we have obtained

$$\|(1 - \chi)u\|_{e^{\omega \cdot} L^2} \leq r(\omega)^{-1} \|\chi'm\|_{e^{\omega \cdot} L^2} \|v\|. \quad (13.6.9)$$

Let us now go from an L^2 estimate of u to a pointwise estimate.

For $t > 0$, which will remain fixed until the end of the argument, let $\chi_+(s) = \widetilde{\chi}(t - s)$ with $\widetilde{\chi}$ as χ above and, in addition, $\operatorname{supp} \widetilde{\chi} \subset]-\infty, t]$, so that $\chi_+(t) = 1$ and $\operatorname{supp} \chi_+ \subset [0, \infty[$. Then

$$(\partial_s - A)(\chi_+(s)u(s)) = \chi_+'(s)u(s)$$

and

$$u(t) = \chi_+(t)u(t) = \int_{-\infty}^t T(t - s)\,\chi_+'(s)\,u(s)\,ds.$$

Hence, we obtain

$$
\begin{aligned}
e^{-\omega t}\|u(t)\| &= e^{-\omega t}\|\chi_+(t)u(t)\| \\
&\leq \int_{-\infty}^t e^{-\omega t}\,m(t - s)\,|\widetilde{\chi}'(t - s)|\,\|u(s)\|\,ds \\
&\leq \int_{-\infty}^t e^{-\omega(t-s)}\,m(t - s)\,|\widetilde{\chi}'(t - s)|\,e^{-\omega s}\,\|u(s)\|\,ds \\
&\leq \|m\widetilde{\chi}'\|_{e^{\omega \cdot} L^2}\,\|u\|_{e^{\omega \cdot} L^2(\operatorname{supp}\chi_+)}. \quad (13.6.10)
\end{aligned}
$$

We assume that

$$\chi = 0 \text{ on } \operatorname{supp}\chi_+. \quad (13.6.11)$$

Then u can be replaced by $(1 - \chi)u$ in the last line of (13.6.10):

$$e^{-\omega t}\|u(t)\| \leq \|m\widetilde{\chi}'\|_{e^{\omega \cdot} L^2}\,\|(1 - \chi)u\|_{e^{\omega \cdot} L^2}.$$

Using (13.6.9), we obtain

$$e^{-\omega t}\|u(t)\| \leq r(\omega)^{-1}\,\|m\chi'\|_{e^{\omega \cdot} L^2}\,\|m\widetilde{\chi}'\|_{e^{\omega \cdot} L^2}\,\|v\|. \quad (13.6.12)$$

Let

$$\operatorname{supp} \chi \subset \,]-\infty, a], \quad \operatorname{supp} \tilde{\chi} \subset \,]-\infty, \tilde{a}], \quad a + \tilde{a} = t, \tag{13.6.13}$$

so that (13.6.11) holds.

Optimal choices for χ and $\tilde{\chi}$ For a given $a > 0$, we look for a χ like that in (13.6.13) such that $\|m\chi'\|_{e^\omega \cdot L^2}$ is as small as possible. By the Cauchy–Schwarz inequality,

$$1 = \int_0^a |\chi'(s)|\, ds \leq \|\chi'm\|_{e^\omega \cdot L^2} \left\| \frac{1}{m} \right\|_{e^{-\omega} \cdot L^2(]0,a[)}, \tag{13.6.14}$$

and so

$$\|\chi'm\|_{e^\omega \cdot L^2} \geq \frac{1}{\|1/m\|_{e^{-\omega} \cdot L^2(]0,a[)}}. \tag{13.6.15}$$

We obtain equality in (13.6.15) if, for some constant C,

$$|\chi'(s)|\, m(s)e^{-\omega s} = C \frac{1}{m(s)} e^{\omega s} \text{ on } [0, a],$$

i.e.,

$$\chi'(s)m(s)e^{-\omega s} = -C \frac{1}{m(s)} e^{\omega s} \text{ on } [0, a], \tag{13.6.16}$$

where C is determined by the condition $1 = \int_0^a |\chi'(s)|\, ds$. We obtain

$$C = \frac{1}{\|1/m\|^2_{e^{-\omega} \cdot L^2(]0,a[)}}.$$

Here we recall that $\chi(s) = 1$ for $s \leq 0$, $\chi(s) = 0$ for $s \geq a$, and there is a unique solution of (13.6.16) given by

$$\chi(s) = C \int_s^a \frac{1}{m(\sigma)^2} e^{2\omega\sigma}\, d\sigma, \ 0 \leq s \leq a.$$

With a similar optimal choice of $\tilde{\chi}$, for which

$$\|\tilde{\chi}'m\|_{e^\omega \cdot L^2} = \frac{1}{\|1/m\|_{e^{-\omega} \cdot L^2(]0,\tilde{a}[)}},$$

we obtain the following from (13.6.12):

$$e^{-\omega t} \|u(t)\| \leq \frac{\|v\|}{r(\omega) \|1/m\|_{e^{-\omega \cdot} L^2(]0,a[)} \|1/m\|_{e^{-\omega \cdot} L^2(]0,\tilde{a}[)}}, \quad (13.6.17)$$

provided that $a, \tilde{a} > 0$ and $a + \tilde{a} = t$ for any $v \in D(A)$. Observing that $D(A)$ is dense in $\mathcal{H}$, we obtain the same estimate for v in $\mathcal{H}$, and this completes the proof of Theorem 13.30.

13.7 Applications: explicit bounds

Theorem 13.30 is based on two assumptions: the existence of some initial control of $\|T(t)\|$ by $m(t)$ and additional information about the resolvent. The existence of $m(t)$ is usually easy but cannot be obtained from abstract analysis. We are in fact looking for explicit bounds. For example, the Hille–Yosida theorem (Theorem 13.22) shows that we can take $m(t) = \widehat{M} \exp \widehat{\omega} t$ for some $\widehat{\omega} \geq \omega$, and in fact we can do so with $\widehat{M} = 1$. We apply Theorem 13.30 with this $m(t)$ and $a = \tilde{a} = t/2$. The term appearing in the denominator of (13.5.10) becomes

$$\left\| \frac{1}{m} \right\|_{e^{-\omega \cdot} L^2(]0,a[)} \left\| \frac{1}{m} \right\|_{e^{-\omega \cdot} L^2(]0,\tilde{a}[)} = \frac{1}{2} \widehat{M}^{-2} t \quad (13.7.1)$$

if $\widehat{\omega} = \omega$, and

$$= \frac{1}{2\widehat{M}^2(\widehat{\omega} - \omega)} \left[1 - \exp((\omega - \widehat{\omega})t) \right] \quad (13.7.2)$$

if $\widehat{\omega} > \omega$.

Hence we obtain an estimate with a new $m(t)$, denoted by $m^{\text{new}}(t)$, where

$$m^{\text{new}}(t) = \frac{2\widehat{M}^2(\widehat{\omega} - \omega)}{r(\omega)[1 - \exp((\omega - \widehat{\omega})t)]} \exp \omega t.$$

This gives in particular the result that $T(t)$ satisfies $P(M, \omega)$, with

$$M = \sup_t \left(\exp -\omega t \, \min(\widehat{M} \exp \widehat{\omega} t, m^{\text{new}}(t)) \right).$$

Let us continue the computation. Without loss of generality, we can assume that $\widehat{\omega} = 0$, and we make the assumptions of Theorem 13.30 for some $\omega < 0$. Combining Theorem 13.30 and the trivial estimate

$$\|T(t)\| \leq \widehat{M} = \widehat{M} \exp -\omega t \exp \omega t,$$

we obtain the result that we have $P(M, \omega)$ with

$$M = \widehat{M} \sup_{t} \left(\min \left(\exp -\omega t, \frac{2\widehat{M} |\omega|}{r(\omega)(1 - \exp \omega t)} \right) \right).$$

This can be rewritten in the form

$$M = \widehat{M} \sup_{u \in]0,1[} \left(\min \left(\frac{1}{u}, \frac{2\widehat{M} |\omega|}{r(\omega)(1 - u)} \right) \right) = 1 + 2\frac{\widehat{M} |\omega|}{r(\omega)}.$$

Proposition 13.31 *Let $T(t)$ be a continuous semigroup such that $P(\widehat{M}, \widehat{\omega})$ is satisfied for some pair $(\widehat{M}, \widehat{\omega})$ and such that $r(\omega) > 0$ for some $\omega < \widehat{\omega}$. Then*

$$\|T(t)\| \le \widehat{M} \left(1 + \frac{2\widehat{M}(\widehat{\omega} - \omega)}{r(\omega)} \right) \exp \omega t. \qquad (13.7.3)$$

In the same spirit, and combining Lemma 13.28 with $\omega' = \omega - sr(\omega)$, we obtain the following extension of (13.7.3):

$$\|T(t)\| \le \widehat{M} \left(\frac{(1 - s)r(\omega) + 2\widehat{M}(\widehat{\omega} - \omega + sr(\omega))}{(1 - s)r(\omega)} \right) \exp ((\omega - sr(\omega))t),$$

$$\forall s \in [0, 1[. \quad (13.7.4)$$

Taking $s = t/(1+t)$ gives a rather optimal decay at ∞ in $\mathcal{O}(t) \exp(\omega - r(\omega))t$.

Remark 13.32 If we now assume instead that the norm of the resolvent on $\operatorname{Re} z \ge 0$ is controlled, and hence we have the case $\omega = \widehat{\omega} = 0$, we obtain

$$\|T(t)\| \le \frac{2\widehat{M}}{r(0)t}.$$

Use of the semigroup property shows that

$$\|T(t)\| \le \left(\frac{2\widehat{M} N}{r(0)t} \right)^N,$$

for any $N \ge 1$. Optimizing over N permits us to deduce an exponential decay of $T(t)$ at ∞.

The limit $\omega \searrow \omega_0$ Consider the situation of Theorem 13.30 and let ω_0 be as in (13.5.5). Assume that $\omega_0 > -\infty$, so that $r(\omega) \to 0$ when $\omega \to \omega_0$. For $t \ge 1$ and $\omega > \omega_0$, we obtain from (13.5.10) the existence of $C > 0$ such that

$$e^{-\omega_0 t} \|T(t)\| \le \frac{e^{t(\omega-\omega_0)}}{r(\omega) \int_0^{1/2} m(s)^{-2} e^{2\omega_0 s} \, ds} \le C \frac{e^{t(\omega-\omega_0)}}{r(\omega)}. \tag{13.7.5}$$

By optimizing over $\omega \in]\omega_0, \omega_0 + \epsilon_0]$, we obtain the existence of C such that

$$e^{-\omega_0 t} \|T(t)\| \le C \exp \Phi(t), \tag{13.7.6}$$

with

$$\Phi(t) = \inf_{\omega \in]\omega_0, \omega_0 + \epsilon_0]} t(\omega - \omega_0) - \ln r(\omega).$$

It is clear that $\lim_{t \to +\infty} \Phi(t)/t = 0$, but to obtain a more quantitative version, we need some information about the behavior of $r(\omega)$ as $\omega \searrow \omega_0$. If, for example,

$$r(\omega) \ge \frac{(\omega - \omega_0)^k}{C} \quad \text{when } 0 < \omega - \omega_0 \le \frac{1}{C},$$

for some constants $C, k > 0$, then by choosing $\omega - \omega_0 = k/t$ in (13.7.5) we obtain, for some constant $\hat{C} > 0$,

$$e^{-\omega_0 t} \|T(t)\| \le \hat{C} t^k, \quad t \ge 1.$$

We can think of a Jordan matrix as the simplest example of this situation.

13.8 Notes

In Section 13.2, we have followed chapter 4 of Trefethen and Embree's book [TrEm]. We were also inspired by chapters 8 and 9 of Brian Davies's book [Dav7] and by remarks in [Dav2, Dav3, Dav4, Dav5]. The concept of the pseudospectrum first appeared in numerical analysis; see Trefethen [Tr1, Tr2]. In some of the literature, the strict inequality $>$ is replaced by $\ge$ in the definition of the pseudospectrum. We have chosen the definition that leads to the simplest statements. The main properties of the pseudospectrum can be found in the above references.

Theorem 13.2 is a weak version of a result by Roch and Silbermann [RoSi]. Numerical examples are treated in [Tr2].

There is a semiclassical version of the pseudospectrum for families A_h. In this case it is possible to relate the ϵ appearing in the definition of the ϵ-pseudospectrum to the parameter h (which is typically in $]0, h_0]$). For example, we may consider $\epsilon(h) = h^N$ (see, for example, [DSZ]).

For Section 13.3, possible references are the books by Dautray and Lions [DaLi, Vol. 5, chapter XVII], Reed and Simon [RS-IV], and Davies [Dav7].

Note that the terminology of Reed and Simon is different. Some other authors use the notion of dissipative operators, with the correspondence that A is accretive if $-A$ is dissipative.

For Section 13.4, the main references are [St], [EnNa1], [Paz], and [Yo]. The basic formula (13.4.5) can be found in [EnNa1, theorem II.1.10]. There is also a presentation in Reed and Simon [RS-II]. As mentioned by F. Nier, Proposition 13.12 is the topic of Problem 52 in [RS-II].

For Section 13.5, proofs of the Gearhart–Prüss theorem can be found in [EnNa1, theorem V.I.11], and in [TrEm, theorem 19.1]; see [Ge, Pr] for the original proof. This theorem also has many other names, including Huang's theorem [HFL] and Greiner's theorem. The proof of the quantitative Gearhart–Prüss theorem is taken from [HeSj2]. As we have seen, the Gearhart–Prüss theorem is a consequence of its quantitative version. For more information, readers can refer to [Dav6].

13.9 Exercises

Exercise 13.1 Let $A \in \mathcal{L}(\mathcal{H})$ and S be invertible in $\mathcal{L}(\mathcal{H})$. Using the quantity $\kappa(S) = \|S\| \, \|S^{-1}\|$, compare the pseudospectra of A and $S^{-1}AS$. Compare $\kappa(S)$ and the instability index introduced in (13.2.3).

Exercise 13.2 Analyze the spectrum and the pseudospectra of $-d^2/dx^2 + d/dx$ on the line.

Exercise 13.3 [Dav5] Determine the spectrum Σ_n and the eigenfunctions of the Dirichlet problem for $-d^2/dx^2 + d/dx$ on $[-n, +n]$. Do these eigenfunctions form a total family in $L^2(]-n, +n[)$? Determine the norm of the projector attached to each eigenvalue (in other words, the instability index of each eigenvalue). What is its behavior as the eigenvalue tends to $+\infty$? Deduce from this that for any $\epsilon > 0$, the ϵ-pseudospectra have no unbounded components.

What is the limit of Σ_n as $n \to +\infty$? Compare this with the result of Exercise 13.2.

Exercise 13.4 [Hag] Let g be a C^∞ function on the circle. Analyze the spectrum and the resolvent of the operator $d/d\theta + g(\theta)$ with domain H^1 on the circle.

Exercise 13.5 [BoKr] Let $\alpha \in \mathbb{R}$. Analyze the spectrum of $-d^2/dx^2$ on the interval $]-1, +1[$ with the boundary condition $u'(\pm 1) = i\alpha u(\pm 1)$. Discuss as a function of α the following questions: Is the operator self-adjoint? Is the spectrum real? Is the operator diagonalizable?

We suggest that you show first that α^2 is an eigenvalue.

Exercise 13.6 [SjZw] Consider on $\mathbb{C}^n$ the matrix $J^{(n)}$ given by $a_{ij}^{(n)} = 0$ except when $j = i + 1$ $(i = 1, \ldots, n - 1)$, in which case $a_{ij}^{(n)} = 1$. Show that the spectrum of $J^{(n)}$ is $\{0\}$. Determine the kernel of $J^{(n)}$.

Consider $e_n = (1, z, \ldots, z^{n-1})$ and compute the norm of $(J^{(n)} - z)e_n$. What can we say about the pseudospectra of $J^{(n)}$? Discuss the dependence on n. Show that for any $\epsilon > 0$ and any $\rho \in]0, 1[$, there exists n_0 such that for $n \geq n_0$, $D(0, \rho) \subset \sigma_\epsilon(J^{(n)})$. Compare this result with the spectrum of τ_r^* in Exercise 6.8.

Exercise 13.7 Show that the unbounded operator $-\Delta + V$ on $L^2(\mathbb{R}^m)$, where V is a C^∞ function such that $\operatorname{Re} V \geq 0$, with domain $C_0^\infty(\mathbb{R}^m)$, is accretive and admits a unique maximally accretive extension.

Hint: Use Theorem 13.14 and the proof of Theorem 9.15.

14

Applications to non-self-adjoint one-dimensional models

Except for the first example, which is something of a toy model, the examples presented in this chapter are strongly motivated by problems from fluid mechanics and superconductivity. It is good to be able to obtain very accurate estimates using the theory of ordinary differential equations, but alternative proofs can be very important when one is attacking problems in many dimensions.

14.1 Analysis of the differentiation operator on an interval

We consider the operator A defined on $L^2(]0, 1[)$ by

$$D(A) = \{u \in H^1(]0, 1[), \; u(1) = 0\}$$

and

$$Au = u', \; \forall u \in D(A).$$

This is clearly a closed operator with a dense domain. The adjoint of A is defined on $L^2(]0, 1[)$ by

$$D(A^*) = \{u \in H^1(]0, 1[), \; u(0) = 0\}$$

and

$$A^*u = -u', \; \forall u \in D(A^*).$$

Lemma 14.1 *With A as above, $\sigma(A) = \emptyset$ and A has a compact resolvent.*

First, we can observe that $(A - z)$ is injective on $D(A)$ for any $z \in \mathbb{C}$. This is simply the observation that $u \in N(A - z)$ satisfies $u'(t) = zu(t)$ and $u(1) = 0$.

It can also be easily verified that for any $z \in \mathbb{C}$, the inverse is given by

$$[(z - A)^{-1}f](x) = \int_x^1 \exp z(x - s) \, f(s) \, ds. \qquad (14.1.1)$$

It is also clear that this operator is compact (for example, because its distribution kernel is in $L^2(]0, 1[\times]0, 1[)$).

Analyzing the ϵ-pseudospectrum is more interesting. For this, we need to estimate the level sets of $\psi(z) := \|(z-A)^{-1}\|$. We recall from Proposition 13.8 that ψ is subharmonic. Observing that for any $\alpha \in \mathbb{R}$, the map $u \mapsto U_\alpha u := \exp i\alpha x \, u$ is a unitary transform on $L^2(]0, 1[)$, which maps $D(A)$ onto $D(A)$ and is such that $U_\alpha^{-1} A U_\alpha = A + i\alpha$, we deduce the following lemma.

Lemma 14.2 ψ *depends only on* Re z. *Moreover, the function* $\mathbb{R} \ni x \mapsto \psi(x)$ *is convex.*

The main result is the following.

Theorem 14.3 *The function* ψ *satisfies*

$$\psi(z) \le \frac{1}{\text{Re } z} \ \text{for Re } z > 0 \qquad (14.1.2)$$

and

$$\psi(z) = -\frac{\exp -\text{Re } z}{2 \, \text{Re } z} + \mathcal{O}\left(\frac{1}{|\text{Re } z|}\right) \text{for Re } z < 0. \qquad (14.1.3)$$

Using also Lemma 14.2, this implies the following for the ϵ-pseudo-spectrum of A.

Corollary 14.4 *For* $\epsilon > 0$, *the* ϵ-*pseudospectrum of* A *is a half-plane of the form*

$$\sigma_\epsilon(A) = \{z \in \mathbb{C} \mid \text{Re } z < c_\epsilon\}, \qquad (14.1.4)$$

with

$$c_\epsilon \sim \begin{cases} (\ln \epsilon) & \text{as } \epsilon \to 0, \\ \epsilon & \text{as } \epsilon \to +\infty. \end{cases} \qquad (14.1.5)$$

According to Lemma 14.2, we can assume that z is real.

Rough estimate from below for $z < 0$ An initial (nonoptimal) approach is to think semiclassically[1] by constructing approximate solutions. Let us take z **real** and consider

$$x \mapsto \phi_z(x) := |2z|^{1/2} \exp zx.$$

This function is not in $N(z - A)$, because it does not satisfy the boundary condition. But when $z \to -\infty$, the boundary condition at $x = 1$ is "almost satisfied." Actually, ϕ_z lives very close to 0. Moreover, for $z < 0$,

$$\|\phi_z\|^2 = 1 - \exp 2z.$$

Hence the norm tends to 1 as $z \to -\infty$.

Let us now consider, for $\eta > 0$, a C^∞ cutoff function χ_η such that $\chi_\eta = 1$ on $[0, 1 - \eta]$ and $\chi_\eta = 0$ on $[1 - \eta/2, 1]$, and introduce

$$\phi_{z,\eta}(x) = \chi_\eta \phi_z.$$

We now observe that $\phi_{z,\eta} \in D(A)$ and that

$$(z - A)\phi_{z,\eta}(x) = -\chi'_\eta \phi_z.$$

The L^2-norm of the right-hand side is exponentially small, like $\exp(1 - \eta)z$.

Returning to the definition of ψ, this shows that for any $\eta > 0$, there exist $z_\eta < 0$ and $C_\eta > 0$ such that

$$\frac{1}{C_\eta} \exp -(1 - \eta)z \le \psi(z) \text{ for } z < z_\eta. \tag{14.1.6}$$

This is not as good as in the statement of the theorem (see (14.1.3)), but it suggests a rather general point of view. We shall complete the analysis of the behavior of ψ as $\operatorname{Re} z \mapsto -\infty$ later.

Rough estimate from above for $z > 0$ Here we shall try to estimate $\psi(z)$ from above by starting from a very simple identity for $(A - z)$ (with z real). For $u \in D(A)$, we have

$$\langle (A - z)u, u \rangle = -z \|u\|^2 + \int_0^1 u'(t) \bar{u}(t) \, dt.$$

---

[1] The semiclassical parameter is $h = 1/|\operatorname{Re} z|$.

Taking the real part, we obtain, for $z \in \mathbb{R}$,

$$- \operatorname{Re} \langle (A - z)u, \, u \rangle = z \|u\|^2 + \frac{1}{2} |u(0)|^2 \geq z \|u\|^2.$$

We then obtain

$$\|(A - z)u\| \geq z \|u\|, \; \forall u \in D(A), \tag{14.1.7}$$

which implies (14.1.2).

Checking of the resolvent using Schur's lemma   Since the operator $(A - z)^{-1}$ is an operator associated with an integral kernel, we can analyze what can be obtained from Schur's lemma (see Lemma 7.1) or from the Hilbert–Schmidt criterion, by computing the L^2-norm of the integral kernel. The kernel is defined (see (14.1.1)) on $]0, 1[\times]0, 1[$ by

$$K(x, s) = \begin{cases} 0 & \text{for } s < x, \\ \exp z(x - s) & \text{for } x < s. \end{cases} \tag{14.1.8}$$

According to Schur's lemma, we have to consider the expressions $\sup_{x \in]0, 1[} \int_0^1 K(x, s) \, ds$ and $\sup_{s \in]0, 1[} \int_0^1 K(x, s) \, dx$, which can be computed explicitly for $z \neq 0$:

$$\sup_{x \in]0, 1[} \int_0^1 K(x, s) \, ds = \frac{1}{z}(1 - e^{-z}) \quad \text{and} \quad \sup_{s \in]0, 1[} \int_0^1 K(x, s) \, dx = \frac{1}{z}(1 - e^{-z}).$$

This gives

$$\psi(z) \leq \frac{1}{z}(1 - e^{-z}). \tag{14.1.9}$$

This is actually an improved version of (14.1.2) for $z > 0$. For $z < 0$, it is better to write (14.1.2) in the form

$$\psi(z) \leq \frac{-1}{z}(e^{-z} - 1), \tag{14.1.10}$$

and to compare this upper bound with the lower bound obtained in (14.1.6).

A more accurate estimate for $z < 0$ We can rewrite $(z - A)^{-1}$ in the form

$$(z - A)^{-1} = R_1 - R_2,$$

with

$$(R_1 v)(x) := \int_0^1 \exp z(x - s) \, v(s) \, ds$$

and

$$(R_2 v)(x) := \int_0^x \exp z(x - s) \, v(s) \, ds.$$

Observing that R_2^* can be treated as in the proof of (14.1.9) (see (14.1.8)), we first obtain

$$||R_2|| = ||R_2^*|| \le -\frac{1}{z}.$$

It remains to check $||R_1||$. This norm can be computed explicitly. R_1 is actually a rank-one operator, and we have

$$||R_1 v|| = || \exp zx || \left| \int_0^1 \exp -zs \, v(s) \, ds \right|.$$

Hence we just have to compute the norm of the linear form

$$v \mapsto \int_0^1 \exp -zs \, v(s) \, ds,$$

which is the L^2-norm of $s \mapsto \exp -zs$. This gives

$$||R_1|| = || \exp z \cdot || \cdot || \exp -z \cdot || = -\frac{1}{2z} \sqrt{(1 - e^{2z})(e^{-2z} - 1)}$$

$$= -\frac{1}{2z} e^{-z} (1 - e^{2z}).$$

Combining the estimates of $||R_1||$ and $||R_2||$ leads to (14.1.3).

Remark 14.5 One can discretize the preceding problem by considering, for $n \in \mathbb{N}^*$, the matrix $A_n = n A_1$ with $A_1 = I + J$, where J is the $n \times n$ matrix such that $J_{i,j} = \delta_{i+1,j}$. It can be observed that the spectrum of A_n is $\{n\}$. It is also interesting to analyze the ϵ-pseudospectra (see Exercise 13.6).

14.2 Another non-self-adjoint operator on the line without a spectrum

We consider the spectrum of the closed operator A associated with the differential operator

$$A_0 = \frac{d}{dx} + x^2 \tag{14.2.1}$$

on the line. We take as the domain $D(A)$ the space of the $u \in L^2(\mathbb{R})$ such that $A_0 u \in L^2(\mathbb{R})$. We note that $C_0^\infty(\mathbb{R})$ is dense for the graph norm in $D(A)$. Hence A is the closed extension of A_0 with domain $C_0^\infty(\mathbb{R})$. We note that this operator is not self-adjoint. The adjoint is the closed extension of $-d/dx + x^2$, with domain $C_0^\infty(\mathbb{R})$.

The following two inequalities are useful. The first is

$$\mathrm{Re}\,\langle Au, u \rangle \geq ||xu||^2 \geq 0, \forall u \in D(A). \tag{14.2.2}$$

This inequality can be proved first for $u \in C_0^\infty(\mathbb{R})$ and then extended to $u \in D(A)$ using the density of $C_0^\infty(\mathbb{R})$. An immediate consequence is that

$$||xu||^2 \leq ||Au||\,||u|| \leq \frac{1}{2}\left(||Au||^2 + ||u||^2\right). \tag{14.2.3}$$

This implies that $D(A)$ (with the graph norm) has continuous injection in the weighted space L_ρ^2, with $\rho(x) = |x|$. Together with the fact that $D(A) \subset H_{\mathrm{loc}}^1(\mathbb{R})$, this implies that $(A + I)$ is invertible and that the inverse is a compact operator (see Section 5.2).

The second inequality is

$$||x^2 u||^2 + ||u'||^2 \leq C\left(||Au||^2 + ||u||^2\right). \tag{14.2.4}$$

To prove this, we observe that

$$||Au||^2 = ||u'||^2 + ||x^2 u||^2 + 2\,\mathrm{Re}\,\langle u', x^2 u \rangle.$$

Using an integration by parts, we obtain

$$-2\,\mathrm{Re}\,\langle u', x^2 u \rangle = 2\,\langle xu, u \rangle.$$

Hence we obtain, using the Cauchy–Schwarz inequality,

$$||Au||^2 \geq ||u'||^2 + ||x^2 u||^2 - 2\,||xu||\,||u||.$$

We can then use (14.2.3) to obtain the conclusion.

Equation (14.2.4) permits us to obtain the following property of the domain of A:

$$D(A) = \{u \in H^1(\mathbb{R}),\ x^2 u \in L^2(\mathbb{R})\}, \tag{14.2.5}$$

which was not a priori obvious.

Proposition 14.6 *A has an inverse, and the inverse is compact. Moreover, its spectrum is empty.*

We consider on $\mathbb{R}$ the differential equation

$$u' + x^2 u = f. \tag{14.2.6}$$

For all $f \in L^2(\mathbb{R})$, we shall show that there exists a unique solution of (14.2.6) in $L^2(\mathbb{R})$. Elementary calculus gives

$$u(x) = e^{-x^3/3} \int_{-\infty}^{x} e^{y^3/3} f(y)\, dy.$$

One has to work a little to show that $u \in L^2$ (this is easier if f is compactly supported). If we denote by $\mathbf{K}$ the operator that associates the solution u with f, the distribution kernel of $\mathbf{K}$ is given by

$$K(x, y) = \begin{cases} 0 & \text{if } y \geq x, \\ e^{(1/3)(y^3 - x^3)} & \text{if } y < x. \end{cases}$$

We note that if there exists a nonzero eigenvalue μ of $\mathbf{K}$ and if u_μ is a corresponding eigenfunction, then u_μ satisfies

$$u_\mu' + x^2 u_\mu = \lambda u_\mu, \quad \text{with } \lambda = \frac{1}{\mu}. \tag{14.2.7}$$

From this, we deduce that $\mathbf{K}$ has no nonzero eigenvalues. It is in fact possible to solve (14.2.7) explicitly, and we find

$$u_\mu(x) = C e^{\lambda x - x^3/3}. \tag{14.2.8}$$

It is then easy to see that no solution of this kind can be in $L^2(\mathbb{R})$ when $C \neq 0$.

To show that $\mathbf{K}$ is compact, we can either use (14.2.5) or show that $\mathbf{K}$ is Hilbert–Schmidt. To prove the second statement, we have to compute the $L^2(\mathbb{R}^2)$-norm of $K(x, y)$. We have

$$\int_{y<x} e^{(2/3)(y^3 - x^3)}\, dx\, dy = \int_{-\infty}^{+\infty} \left(\int_{-\infty}^{x} e^{(2/3)(y^3 - x^3)}\, dy \right) dx.$$

Dividing the domain of integration into two parts, we first consider

$$\int_{-1}^{+1} \left(\int_{-\infty}^{x} e^{(2/3)(y^3 - x^3)}\, dy \right) dx,$$

which is bounded very roughly from above by

$$2e \int_{-\infty}^{1} e^{(2/3)y^3} \, dy.$$

Then we look at

$$\int_{|x|>1} \left(\int_{-\infty}^{x} e^{(2/3)(y^3 - x^3)} \, dy \right) dx.$$

Here we observe that

$$e^{(2/3)(y^3 - x^3)} = e^{(2/3)(y-x)(y^2 + yx + x^2)},$$

and that

$$(y^2 + yx + x^2) \geq \frac{1}{2}(y^2 + x^2) \geq \frac{1}{2}x^2.$$

This leads, as $y \leq x$, to the upper bound

$$e^{(2/3)(y^3 - x^3)} \leq e^{(1/3)(y-x)x^2},$$

and to

$$\int_{|x|>1} \left(\int_{-\infty}^{x} e^{(2/3)(y^3 - x^3)} \, dy \right) dx \leq \int_{|x|>1} \left(\int_{-\infty}^{x} e^{(1/3)(y-x)x^2} \, dy \right) dx$$

$$\leq 3 \int_{|x|>1} \frac{1}{x^2} \, dx \, < +\infty. \qquad (14.2.9)$$

So **K** is Hilbert–Schmidt, and hence compact.

Non-self-adjoint effects   We can also try to estimate $\mathbf{K}_\lambda = (A - \lambda)^{-1}$. First, we observe that we can reduce the computation to the case where λ is real. We have in fact

$$\mathbf{K}_\lambda = e^{-i\operatorname{Im}\lambda x} \, \mathbf{K}_{\operatorname{Re}\lambda} \, e^{i\operatorname{Im}\lambda x},$$

which shows in addition that $\mathbf{K}_\lambda$ is unitarily equivalent to $\mathbf{K}_{\operatorname{Re}\lambda}$.

We now assume that λ is real. For $\lambda < 0$, observing that

$$\operatorname{Re}\langle u' + x^2 u - \lambda u, \, u \rangle_{L^2} = \langle x^2 u - \lambda u, \, u \rangle \geq -\lambda \|u\|^2,$$

we easily find the estimate

$$\|\mathbf{K}_\lambda\|_{\mathcal{L}(L^2)} \leq -\frac{1}{\lambda}.$$

The case $\lambda = 0$ has been treated before. Observing that $\mathbf{K}_\lambda$ is holomorphic over $\mathbb{C}$ by Proposition 14.6, we obtain the existence of a constant C such that

$$||\mathbf{K}_\lambda||_{\mathcal{L}(L^2)} \leq C, \ \forall \lambda \in [-1, 1].$$

So, we have to consider the case where $\lambda \geq 1$ and check the estimate as $\lambda \rightarrow +\infty$.

Proceeding as in the case $\lambda = 0$, we first obtain

$$K_\lambda(x, y) = \begin{cases} 0 & \text{if } y \geq x, \\ e^{((1/3)(y^3 - x^3) - \lambda(y-x))} & \text{if } y < x. \end{cases} \tag{14.2.10}$$

Again we see that $K_\lambda(x, y)$ is in $L^2(\mathbb{R}^2)$:

$$||K_\lambda||^2_{L^2(\mathbb{R}^2)} = \int_{-\infty}^{+\infty} \left(\int_{-\infty}^{x} e^{((2/3)(y^3 - x^3) - 2\lambda(y-x))} \, dy \right) dx. \tag{14.2.11}$$

The proof is now similar to the case $\lambda = 0$. We cut the domain of integration into three subdomains in the variable x delimited by the two points $-3\sqrt{\lambda}$ and $3\sqrt{\lambda}$.

Subdomains $]-\infty, -3\sqrt{\lambda}[$ **and** $]3\sqrt{\lambda}, +\infty[$ We have to estimate

$$\int_{|x| \geq 3\sqrt{\lambda}} \left(\int_{-\infty}^{x} e^{((2/3)(y^3 - x^3) - 2\lambda(y-x))} \, dy \right) dx.$$

As before (in the case $\lambda = 0$), we observe that

$$e^{(2/3)(y^3 - x^3)} = e^{(2/3)(y-x)(y^2 + yx + x^2)}.$$

This leads to the upper bound

$$e^{(2/3)(y^3 - x^3)} e^{-2\lambda(y-x)} \leq e^{(1/3)(y-x)(x^2 - 6\lambda)}.$$

By simple integration, we obtain

$$\int_{|x| > 3\sqrt{\lambda}} \left(\int_{-\infty}^{x} e^{((2/3)(y^3 - x^3) - 2\lambda(y-x))} \, dy \right) dx$$

$$\leq 3 \int_{|x| > 3\sqrt{\lambda}} \frac{1}{x^2 - 6\lambda} \, dx < 27 \int_{|x| > 1} \frac{1}{x^2} \, dx < +\infty.$$

The last bound is independent of λ.

Subdomain $]-3\sqrt{\lambda}, 3\sqrt{\lambda}[$ The integral

$$\int_{-3\sqrt{\lambda}}^{3\sqrt{\lambda}} \left(\int_{-\infty}^{x} e^{((2/3)(y^3-x^3)-2\lambda(y-x))} \, dy \right) dx$$

is bounded by

$$3\lambda^{-1/2} e^{(4/3)(3\lambda^{1/2})^3}.$$

Of course, this is very rough (we shall obtain a much better bound in the next section).

Returning to complex λ's, we have finally proved the following result.

Proposition 14.7 *There exist* $C > 0$, $C_0 > 0$, *and* $C_1 > 0$ *such that for* $\operatorname{Re} \lambda \geq 1$, *we have*

$$\|\mathbf{K}_\lambda\|_{\mathrm{HS}} \leq C \, |\operatorname{Re} \lambda|^{C_0} \, e^{C_1 \, |\operatorname{Re} \lambda|^{3/2}}. \tag{14.2.12}$$

The following theorem is useful for a more accurate asymptotic analysis.

Theorem 14.8 *(Laplace's method.) Let* ϕ *be a function in* $C^\infty(\overline{D}(0, 1))$, *where* $\overline{D}(0, 1)$ *is the closed unit ball in* $\mathbb{R}^n$ *such that*

- $\phi \geq 0$ *on* $B(0, 1)$, $\phi > 0$ *on* $\partial B(0, 1)$;
- $\phi(0) = \nabla\phi(0) = 0$;
- ϕ *has, in* $D(0, 1)$, *a unique nondegenerate minimum at* 0.

Let a *be a function in* $C^\infty(\overline{D}(0, 1))$, *and consider the integral*

$$I(a, \phi; h) = \int_{B(0,1)} a(x) \exp\left(-\frac{\phi(x)}{h} \right) dx, \tag{14.2.13}$$

for $h \in]0, h_0]$. *Then* $I(a, \phi; h)$ *has, as* $h \to 0$, *the following asymptotic behavior:*

$$I(a, \phi; h) \sim h^{n/2} \sum_j \alpha_j h^j, \tag{14.2.14}$$

with

$$\alpha_0 = (2\pi)^{n/2} a(0) (\det(\operatorname{Hess}\phi)(0))^{-1/2}. \tag{14.2.15}$$

This theorem is used in the following remark.

Remark 14.9 Using the Laplace method and a dilation in the integral defining the square of $||\mathbf{K}_\lambda||^2_{\mathrm{HS}}$ (see (14.2.11)), we have to look for the minimum of the phase ϕ, where

$$(x, y) \mapsto \phi(x, y) := -\frac{2}{3}(y^3 - x^3) + 2(y - x),$$

on $\{y \le x\}$. This is at $(x, y) = (1, -1)$, with $1/h = \lambda^{3/2}$. The asymptotics of $||\mathbf{K}_\lambda||_{\mathrm{HS}}$ given in (14.2.12) can be improved by making the optimal choice $C_1 = \frac{4}{3}$. Note that $||\mathbf{K}_\lambda||_{\mathrm{HS}} \ge ||\mathbf{K}_\lambda||_{\mathcal{L}(L^2)}$.

One can also find a lower bound for $||\mathbf{K}_\lambda||_{\mathcal{L}(L^2)}$ using approximate eigenfunctions of $d/dx + x^2 - \lambda$ (see (14.2.8)).

Remark 14.10 Using the Fourier transform, it can be seen that the operator is isospectral to the complex Airy operator

$$D_x^2 + ix \tag{14.2.16}$$

(i.e., it has the same spectrum). This operator will be analyzed in Section 14.3. The results of this section permit us to improve the estimate (14.2.12) on $||\mathbf{K}_\lambda||_{\mathcal{L}(L^2)}$ (see Theorem 14.11).

14.3 The complex Airy operator on $\mathbb{R}$

14.3.1 Definition and main properties

Bearing in mind Remark 14.10, we can just lift the properties obtained for the operator $d/dx + x^2$ in the previous section to the operator $D_x^2 + ix$ with $D_x = -i\, d/dx$, using the Fourier transform. But let us see what we can obtain by a direct approach.

The complex Airy operator can be defined by starting from the differential operator on $C_0^\infty(\mathbb{R})$ $\mathcal{A}_0^+ := D_x^2 + ix$. Using the inequality

$$\mathrm{Re}\,\langle \mathcal{A}_0 u, u \rangle = ||u'||^2 \ge 0, \tag{14.3.1}$$

we can see that $\mathcal{A}_0$ is accretive, and hence closable, and that the domain of its closure $\mathcal{A} = \overline{\mathcal{A}_0}$ is

$$D(\mathcal{A}) = \{u \in H^1(\mathbb{R}),\ (D_x^2 + ix)\,u \in L^2(\mathbb{R})\}.$$

Following the proof of (14.2.4) or using the Fourier transform, we can then show that we have

$$D(\mathcal{A}) = \{u \in H^2(\mathbb{R}),\ x\,u \in L^2(\mathbb{R})\}. \tag{14.3.2}$$

In particular, $\mathcal{A}$ has a compact resolvent, and

$$\mathrm{Re}\,\langle \mathcal{A}u,\ u\rangle \geq 0. \tag{14.3.3}$$

It can also be shown that $\mathcal{A}$ is maximally accretive. We can, for example, use Theorem 13.14. In the present case, it can be immediately verified that $D_x^2 - ix + \lambda$ is injective in $\mathcal{S}'(\mathbb{R})$ (by taking the Fourier transform). Using Corollary 13.24, we obtain the result that $-\mathcal{A}$ is the generator of a contraction semigroup S_t.

A very special property of this operator is that for any $a \in \mathbb{R}$,

$$T_a \mathcal{A} = (\mathcal{A} - ia)T_a, \tag{14.3.4}$$

where T_a is the translation operator $(T_a u)(x) = u(x - a)$, which is of course a unitary operator. As an immediate consequence, we obtain the result that the spectrum is empty and that the resolvent of $\mathcal{A}$, which is defined for any $\lambda \in \mathbb{C}$, satisfies

$$\|(\mathcal{A} - \lambda)^{-1}\| = \|(\mathcal{A} - \mathrm{Re}\,\lambda)^{-1}\|. \tag{14.3.5}$$

The most interesting property is the control of the resolvent for $\mathrm{Re}\,\lambda \geq 0$.

Proposition 14.11 *There exist two positive constants C_1 and C_2, such that*

$$C_1\,|\mathrm{Re}\,\lambda|^{-1/4}\exp\frac{4}{3}|\mathrm{Re}\,\lambda|^{3/2} \leq \|(\mathcal{A}-\lambda)^{-1}\| \leq C_2\,|\mathrm{Re}\,\lambda|^{-1/4}\exp\frac{4}{3}|\mathrm{Re}\,\lambda|^{3/2}. \tag{14.3.6}$$

14.3.2 Return to the previous model

The proof of the (rather standard) upper bound is based on a direct analysis of the semigroup in the Fourier representation. We note that

$$\mathcal{F}(D_x^2 + i\,x)\mathcal{F}^{-1} = \xi^2 - \frac{d}{d\xi}. \tag{14.3.7}$$

Semigroup approach   There is a need to improve on what we have done in the previous section.

Using the method of characteristics for evolution problems in the Fourier representation, we can easily obtain

$$\mathcal{F}S_t\mathcal{F}^{-1}v = \exp\left(-\xi^2 t - \xi t^2 - \frac{t^3}{3}\right)v(\xi + t), \tag{14.3.8}$$

and this implies immediately

$$\|S_t\| = \exp \max_{\xi} \left(-\xi^2 t - \xi t^2 - \frac{t^3}{3} \right) = \exp \left(-\frac{t^3}{12} \right). \tag{14.3.9}$$

Note that this decay implies, by the Gearhart–Prüss theorem (see Theorem 13.27 and the discussion there), that the spectrum is empty.

We can then obtain an estimate of the resolvent by using, for $\lambda \in \mathbb{C}$, the formula

$$(\mathcal{A} - \lambda)^{-1} = \int_0^{+\infty} \exp -t(\mathcal{A} - \lambda)\, dt. \tag{14.3.10}$$

For a closed accretive operator, (14.3.10) is standard when $\operatorname{Re} \lambda < 0$, but the estimate (14.3.9) of S_t immediately gives a holomorphic extension of the right-hand side to the whole space, giving for $\lambda > 0$ the estimate

$$\|(\mathcal{A} - \lambda)^{-1}\| \leq \int_0^{+\infty} \exp \left(\lambda t - \frac{t^3}{12} \right) dt. \tag{14.3.11}$$

The asymptotic behavior of this integral as $\lambda \to +\infty$ can be immediately obtained by using the Laplace method and the dilation $t = \lambda^{1/2} s$ in the integral.

In the present case, everything can be computed explicitly. We observe that

$$\int_0^{+\infty} \exp \left(-\xi^2 t - \xi t^2 - \frac{t^3}{3} \right) e^{\lambda t} v(\xi + t)\, dt$$
$$= \int_\xi^{+\infty} \exp \left(\frac{\xi^3 - s^3}{3} + \lambda(s - \xi) \right) v(s)\, ds, \tag{14.3.12}$$

which gives effectively the expression for $(\mathcal{A} - \lambda)^{-1}$ (see (14.2.10), which gives the result for the adjoint).

Quasimodes   The proof of the lower bound is obtained by constructing quasi-modes for the operator $(\mathcal{A} - \lambda)$ in its Fourier representation. We observe (assuming $\lambda > 0$) that

$$\xi \mapsto u(\xi; \lambda) := \exp \left(-\frac{\xi^3}{3} + \lambda \xi - \frac{2}{3} \lambda^{3/2} \right) \tag{14.3.13}$$

is a solution of

$$\left(\frac{d}{d\xi} + \xi^2 - \lambda \right) u(\xi; \lambda) = 0. \tag{14.3.14}$$

By multiplying $u(\cdot; \lambda)$ by a cutoff function χ_λ with support in $]-\sqrt{\lambda}, +\infty[$ and where $\chi_\lambda = 1$ on $]-\sqrt{\lambda} + 1, +\infty[$, we obtain a very good quasimode, concentrated as $\lambda \to +\infty$ around $\sqrt{\lambda}$, with an error term giving almost[2] the announced lower bound for the resolvent.

14.4 The complex harmonic oscillator

Another interesting example to analyze in the same spirit is the complex harmonic oscillator

$$H_+ := -\frac{d^2}{dx^2} + ix^2,$$

initially defined on $C_0^\infty(\mathbb{R})$, sometimes also called the Davies operator.

We immediately have

$$\langle H_+ u, u \rangle \geq ||u'||^2, \ \forall u \in C_0^\infty(\mathbb{R}). \tag{14.4.1}$$

We can see immediately from (14.4.1) that this operator is accretive, and hence closable. We denote its closure by $\overline{H_+}$. Moreover, as seen in Exercise 13.7, the operator is maximally accretive, and we deduce from the definition and (14.4.1) that its domain is

$$D(\overline{H_+}) = \{u \in H^1(\mathbb{R}), \ xu \in L^2(\mathbb{R}), \ (D_x^2 + ix^2)u \in L^2(\mathbb{R})\}. \tag{14.4.2}$$

This implies that $\overline{H_+}$ has a compact resolvent.

Moreover, it can be shown (see the proof of (14.2.4) together with (14.4.1)) that

$$D(\overline{H_+}) = \{u \in H^2(\mathbb{R}), \ xu \in H^1(\mathbb{R}), \ x^2u \in L^2(\mathbb{R})\}. \tag{14.4.3}$$

We can determine the spectrum of this very simple operator explicitly. The spectrum can be seen as the spectrum of $-d^2/dx^2 + x^2$ rotated by $\pi/4$ in the complex plane.

Proposition 14.12

$$\sigma(\overline{H_+}) = \{e^{i\pi/4}(2n - 1), \ n \geq 1\}. \tag{14.4.4}$$

Proof This has two steps.

The inclusion is easy when we observe that the functions $\varphi_n(x) := \phi_n(e^{i\pi/8}x)$, with ϕ_n as defined in (1.3.8), are eigenfunctions.

---

[2] The cutoff must be improved to obtain an optimal result.

To show the equality, let λ be an eigenvalue and suppose that there exists $u \in D(\overline{H_+})$ such that $u \neq 0$ and

$$\overline{H_+}u = \lambda u.$$

Taking the scalar product with $\bar{\varphi}_n$, we obtain the result that

$$(\lambda - \lambda_n)\langle u, \bar{\varphi}_n \rangle = 0, \quad \text{with } \lambda = e^{i\pi/4}(2n - 1).$$

But the vector space generated by the $\bar{\varphi}_n$ is dense in $L^2(\mathbb{R})$. Hence, if $u \neq 0$, this implies the existence of n such that $\lambda = \lambda_n$. $\square$

Let us now sketch another proof, which does not involve an explicit knowledge of the eigenfunctions.

Analytic dilation   If we perform a dilation $x = \rho y$, the operator $-d^2/dx^2 + ix^2$ becomes the operator $-\rho^{-2} d^2/dy^2 + i\rho y^2$, which is unitarily equivalent. We can now consider the family of operators

$$\rho \mapsto H_\rho := -\rho^{-2} \frac{d^2}{dx^2} + i\rho^2 x^2,$$

for ρ in a sector in $\mathbb{C}$.

It can be shown that for ρ in (an open neighborhood of) the sector $\arg \rho \in]-3\pi/16, \pi/16[$, the spectrum is discrete and depends holomorphically on ρ (this comes from Kato's theory). But for ρ real, the eigenvalues are constant. Hence they should be constant in the whole sector. This is a particular case of the so-called Combes–Thomas argument. Taking $\rho = e^{-i\pi/8}$, we obtain

$$H_{e^{-\pi/8}} = e^{i\pi/4} \left(-\frac{d^2}{dx^2} + x^2 \right),$$

and recover (14.4.4).

This operator is maximally accretive, and we can apply Theorem 13.26 with $A = -\overline{H_+}$. We shall show that for fixed $\operatorname{Re} z$ as $\operatorname{Im} z \to +\infty$,

$$\lim_{\operatorname{Im} z \to +\infty} \|(\overline{H_+} - z)^{-1}\| = 0.$$

More precisely, we have the following result.

Proposition 14.13   *For any compact interval K, there exists $C > 0$ such that*

$$\|(\overline{H_+} - z)^{-1}\| \leq C |\operatorname{Im} z|^{-1/3}, \quad \text{for } \operatorname{Im} z \geq C, \operatorname{Re} z \in K.$$

Proof It is enough to check the norm of $(D_x^2 + ix^2 - it)^{-1}$ as $t \to +\infty$ and to show that there exist C and $t_0 \geq 2$ such that for $t \geq t_0$,

$$||(D_x^2 + ix^2 - it)^{-1}||_{\mathcal{L}(L^2)} \leq C |t|^{-1/3}. \tag{14.4.5}$$

Hence, with $P_t = (D_x^2 + ix^2 - it)$, we have to show that for $t \geq t_0$,

$$t^{2/3} ||u||^2 \leq C ||P_t u||^2, \ \forall u \in C_0^\infty(\mathbb{R}). \tag{14.4.6}$$

It is enough to prove that, for $t \geq t_0$,

$$t^{2/3} ||u||^2 \leq C (||P_t u||^2 + ||u||^2), \ \forall u \in C_0^\infty(\mathbb{R}). \tag{14.4.7}$$

First, considering $\text{Re} \langle P_t u, u \rangle$, we observe that

$$||u'||^2 \leq (||P_t u||^2 + ||u||^2). \tag{14.4.8}$$

Step 1. We now check for functions with support outside $\pm\sqrt{t}$. For this, we introduce a regularized version $\chi_t(x)$ of the function $\text{sign}\,(x^2 - t)$ (see (7.2.3) for the definition of this function). This function satisfies

$$\chi_t(x)(x^2 - t) \geq 0,$$

and, for some $\alpha > \frac{1}{3}$,

$$\chi_t(x) = \text{sign}\,(x^2 - t) \text{ if } |x \pm \sqrt{t}| \geq t^\alpha.$$

Moreover, χ_t satisfies

$$|\chi_t'(x)| \leq C t^{-\alpha}.$$

We then analyze $\text{Im}\,\langle \chi_t P_t u, u \rangle$ and write

$$\text{Im}\,\langle \chi_t P_t u, u \rangle = \int \chi_t(x)(x^2 - t)\,|u(x)|^2\,dx + \text{Im} \int \chi_t'(x)u'(x)\,\bar{u}(x)\,dx.$$

Hence

$$\int \chi_t(x)(x^2 - t)\,|u(x)|^2\,dx \leq C\,(||P_t u||^2 + ||u||^2 + t^{-\alpha}\,||u'||\,||u||)$$

$$\leq 2C\,(||P_t u||^2 + ||u||^2). \tag{14.4.9}$$

This corresponds to checking outside the points $\pm\sqrt{t}$ if we observe that when $\inf(|x - t^{1/2}|, |x + t^{1/2}) \geq t^{\alpha}$ and $t \geq 2$, then $|x^2 - t| \geq t^{1/2+\alpha} \geq t^{2/3}$.

Step 2. Close to $\pm\sqrt{t}$, we approximate P_t by $D_x^2 + 2i\sqrt{t}(x \mp \sqrt{t})$. For this, we introduce two functions $\phi_{\pm,t}$ localized around each point $\pm\sqrt{t}$ (say in $] \pm t^{1/2} - 2t^{\alpha}, \pm t^{1/2} + 2t^{\alpha}[$). We note that $\phi'_{\pm,t}$ and $\phi''_{\pm,t}$ are bounded with respect to t and x. Then we observe that

$$P_t\phi_{+,t}u = \phi_{+,t}P_tu - \phi''_{+,t}u - 2\phi'_{+,t}u'.$$

The norm of the right-hand side is controlled by $C\left(\|P_tu\| + \|u\|\right)$:

$$\|P_t\phi_{+,t}u\| \leq C\left(\|P_tu\| + \|u\|\right). \tag{14.4.10}$$

Here, we use an approximation by the complex Airy operator and recall the following property (after a dilation):

$$t^{1/3}\|v\| + t^{1/2}\|xv\| \leq C\|-v'' + 2i\sqrt{t}\,xv\| \tag{14.4.11}$$

(see (14.2.4) and Proposition 14.6 after a Fourier transform). Changing x to $x - \sqrt{t}$ and applying this inequality to $\phi_{+,t}u$, it can be seen immediately by comparison that

$$t^{1/3}\|\phi_{+,t}u\| + t^{1/2}\|(x - \sqrt{t})\phi_{+,t}u\| \leq C(\|P_t(\phi_{+,t}u)\| + \|(x-\sqrt{t})^2\phi_{+,t}u\|)$$
$$\leq \widehat{C}\left(\|P_t(\phi_{+,t}u)\| + 2t^{\alpha}\|(x - \sqrt{t})\phi_{+,t}u\|\right).$$

This implies, if $\alpha < \frac{1}{2}$ and t is large enough,

$$t^{1/3}\|\phi_{+,t}u\| \leq C\|P_t(\phi_{+,t}u)\| \leq \tilde{C}(\|P_tu\| + \|u\|). \tag{14.4.12}$$

Similarly, we obtain

$$t^{1/3}\|\phi_{-,t}u\| \leq \tilde{C}\left(\|P_tu\| + \|u\|\right). \tag{14.4.13}$$

Using (14.4.9), (14.4.12), and (14.4.13), we obtain (14.4.5) for t large enough. $\qquad\square$

Remark 14.14 As $\mathrm{Im}\,z \to -\infty$, we have, by more elementary estimates,

$$\|(\overline{H_+} - z)^{-1}\| \leq |\mathrm{Im}\,z|^{-1} \text{ for } \mathrm{Im}\,z < 0.$$

This estimate is far from optimal for Im z close to 0. If we observe that

$$\mathcal{F}(\overline{H_+} - z)\mathcal{F}^{-1} = \left(\xi^2 - i\frac{d^2}{d\xi^2} - z\right) = i\left(-\frac{d^2}{d\xi^2} - i\xi^2 + iz\right),$$

we obtain the result that

$$\|(\overline{H_+} - z)^{-1}\| = \|(\overline{H_+} - i\bar{z})^{-1}\|.$$

This implies a symmetry of $\psi(z)$ in accordance with the map $z \mapsto i\bar{z}$, i.e., $(x, y) \mapsto (y, x)$. This can be observed in Figure 14.1, which shows the level lines of ψ.

Remark 14.15 H_+^* is the closure of $H_- := -d^2/dx^2 - ix^2$ with domain $C_0^\infty(\mathbb{R})$. It is clear that H_+^* is an extension of H_-, and hence of its closure. But, like $\overline{H_+}$, the closure $\overline{H_-}$ of H_- is surjective. Now, if $u \in D(H_+^*)$, there exists $v \in D(\overline{H_-})$ such that $\overline{H_-}v = H_+^*u$. But H_+^* is injective, and this implies $u = v$ and $u \in D(\overline{H_-})$.

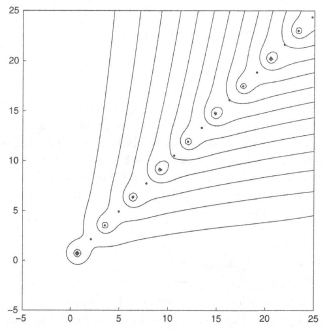

Figure 14.1 Pseudospectra of the Davies operator. Reproduced with permission from [Dav7].

Remark 14.16 This example shows that when operators are not self-adjoint, many things can occur. In particular, the result that if $z \in \rho(T)$ then $\|(T - z)^{-1}\|$ is controlled by $1/d(\lambda, \sigma(T))$ becomes wrong.

14.5 The complex Airy operator on $\mathbb{R}^+$

14.5.1 The (real) Airy operator on $\mathbb{R}^+$

We start from the sesquilinear form

$$(u, v) \mapsto \int_0^{+\infty} u'(t)\,\overline{v'(t)}\,dt + \int_0^{+\infty} t\,u(t)\,\overline{v(t)}\,dt + \int_0^{+\infty} u(t)\,\overline{v(t)}\,dt$$

on $C_0^\infty(\mathbb{R}^+) \times C_0^\infty(\mathbb{R}^+)$ and use the Lax–Milgram theorem (Theorem 3.6). We obtain an unbounded operator S_0 on $L^2(\mathbb{R}^+)$, whose domain is

$$D(S_0) = \{u \in H_0^1(\mathbb{R}^+), t^{1/2}u \in L^2(\mathbb{R}^+), (D_t^2 + t)u \in L^2(\mathbb{R}^+)\}, \quad (14.5.1)$$

and which is defined (in the sense of distributions) for $u \in D(S_0)$ by

$$S_0 u = (D_t^2 + t + 1)u. \quad (14.5.2)$$

Moreover, S_0 is injective on $D(S_0)$.

The Airy operator $\mathcal{A}_D^{\mathbb{R}}$ is defined as $S_0 - 1$, and it is easy to see that it is injective and self-adjoint and has a compact resolvent. Hence the spectrum is discrete and consists of a sequence of real eigenvalues μ_j tending to $+\infty$. The corresponding eigenfunctions can be obtained in the following way. We start from the Airy function $A(x)$, which can be defined as the inverse Fourier transform of $ce^{i\xi^3/3}$ and is the unique solution in $\mathcal{S}'$ of $D_x^2 + x$ on the line. It can be shown that if c is chosen such that $A(0) = 1$, the function[3] $A(x)$ has the following properties:

- $A(x) > 0$ for $x \geq 0$.
- $A(x)$ is exponentially decreasing at $+\infty$.
- There exists a sequence of α_j tending to $-\infty$ such that $A(\alpha_j) = 0$.
- $A(x)$ admits a holomorphic extension in a complex sector $\Sigma_{]-\omega,\omega[} = \{z \in \mathbb{C}, |\arg z| < \omega\}$ (with $\omega = \pi/3$), which is a rapidly decreasing function along each half-line starting from the origin and contained in $\Sigma_{]-\omega,\omega[}$.

---

[3] A more usual notation is $\mathrm{Ai}(x)$.

More precisely (although this is not needed), we have, for some nonzero constant c,

$$A(z) \sim \frac{c}{2} z^{-1/4} e^{-(2/3)z^{3/2}} \quad \text{for} \ |\arg z| < \frac{\pi}{3}, \tag{14.5.3}$$

and

$$A(-z) \sim c\, z^{-1/4} \sin\left(\frac{2}{3}z^{3/2} + \frac{\pi}{4}\right) \quad \text{for} \ |\arg z| < \frac{\pi}{3}, \tag{14.5.4}$$

as $|z| \to +\infty$.

The standard Airy function is normalized with $c = 1/\sqrt{\pi}$.

Having these properties, it is easy to verify that

$$\psi_j(x) = A(x + \alpha_j)$$

($x \geq 0$) is an eigenfunction corresponding to the eigenvalue $\mu_j = -\alpha_j$, and we can assume that any eigenfunction is of this form. Using the spectral theorem, we know that the ψ_j form an orthonormal basis after renormalization. Hence we have the following result.

Proposition 14.17 *The (real) Airy operator on $\mathbb{R}^+$ with the Dirichlet condition $\mathcal{A}_D^{\mathbb{R}}$ is self-adjoint, has a compact resolvent, and admits an orthonormal basis ψ_j associated with the sequence*

$$\mu_j = -\alpha_j, \tag{14.5.5}$$

where the α_j's are the ordered zeros (in decreasing order) of the Airy function. Moreover,

$$0 < \mu_1 < \cdots < \mu_j < \mu_{j+1} < \cdots$$

and

$$\lim_{j \to +\infty} \mu_j = +\infty.$$

14.5.2 The complex Airy operator on $\mathbb{R}^+$: definition and spectrum

We now consider the Airy operator on the half-line. We start from the sesquilinear form

$$(u, v) \mapsto \int_0^{+\infty} u'(x)\, \overline{v'(x)}\, dx + i \int_0^{+\infty} x\, u(x)\, \overline{v(x)}\, dx + \int_0^{+\infty} u(x)\, \overline{v(x)}\, dx$$

on $C_0^\infty(\mathbb{R}^+) \times C_0^\infty(\mathbb{R}^+)$ and use the Lax–Milgram theorem (Theorem 3.6). We obtain an unbounded operator S on $L^2(\mathbb{R}^+)$, whose domain is

$$D(S) = \{u \in H_0^1(\mathbb{R}^+), x^{1/2}u \in L^2(\mathbb{R}^+), (D_x^2 + i\,x)u \in L^2(\mathbb{R}^+)\}, \quad (14.5.6)$$

and which is defined (in the sense of distributions) by

$$Su = (D_x^2 + i\,x + 1)u. \tag{14.5.7}$$

Moreover, S is injective on $D(S)$.

We then define the Dirichlet realization of the Airy operator $D_x^2 + ix$ as

$$\mathcal{A}_D := S - 1.$$

By construction, we have

$$\mathrm{Re}\,\langle \mathcal{A}^D u,\, u \rangle \geq 0, \; \forall u \in D(\mathcal{A}^D). \tag{14.5.8}$$

Theorem 13.14 shows that $\mathcal{A}^D$ is maximally accretive and, by Corollary 13.24, $-\mathcal{A}^D$ is the generator of a contraction semigroup. Moreover, the operator has a compact inverse, and hence the spectrum (if any) is discrete and contained in $\{\mathrm{Re}\,\lambda > 0\}$.

Let us now define

$$\phi_j(x) = \psi_j(xe^{i\pi/6}), \; j \in \mathbb{N}.$$

From the properties presented in the preceding section, ψ_j is well defined in $D(\mathcal{A}^D)$ and satisfies

$$\mathcal{A}^D \phi_j = \lambda_j \phi_j, \tag{14.5.9}$$

where

$$\lambda_j = e^{i\pi/3} \mu_j, \tag{14.5.10}$$

with μ_j defined in (14.5.5). Hence we obtain the result that the spectrum of $\mathcal{A}^D$ $\sigma(\mathcal{A}^D)$ contains these eigenvalues as a subset:

$$\cup_{j=1}^{+\infty}\{\lambda_j\} \subset \sigma(\mathcal{A}^D).$$

We now give a hint as to how we can prove that we have exhausted the set of eigenvalues. Let u be an eigenfunction of the complex Airy operator corresponding to some λ. It is easy to see that it extends as a holomorphic function in $\mathbb{C}\backslash 0$. It is more difficult (requiring Sibuya's theory) to show that u extends as a rapidly decreasing function along each half-line in a sector $\Sigma_{]-\pi/3,\pi/6[}$. We can then show that $x \mapsto u(e^{-i\pi/6}x)$ is an eigenfunction of the real Airy operator with corresponding eigenvalue $\lambda e^{-i\pi/3}$. Hence there should be some μ_j such that $\lambda e^{-i\pi/3} = \mu_j$. To sum up, we have proved the following proposition.

Proposition 14.18 $\mathcal{A}^D$ *is maximally accretive, with a compact resolvent, and its spectrum is given by*

$$\sigma(\mathcal{A}^D) = \cup_{j=1}^{+\infty}\{\lambda_j\}, \tag{14.5.11}$$

with λ_j as defined in (14.5.10).

The following property can also be shown.

Proposition 14.19 *The vector space generated by the eigenfunctions of the Dirichlet realization of the complex Airy operator in $\mathbb{R}^+$ is dense in $L^2(\mathbb{R}^+)$.*

14.5.3 Estimates of the resolvent for the complex Airy operator on $\mathbb{R}^+$

We now arrive at an analysis of the properties of the semigroup and an estimate of the resolvent.

As before, we have, for $\mathrm{Re}\,\lambda < 0$,

$$\|(\mathcal{A}^D - \lambda)^{-1}\| \leq \frac{1}{|\mathrm{Re}\,\lambda|}. \tag{14.5.12}$$

If $\mathrm{Im}\,\lambda < 0$, we obtain a similar inequality, so the main question remaining is the analysis of the resolvent in the set $\mathrm{Re}\,\lambda \geq 0$, $\mathrm{Im}\,\lambda \geq 0$, which corresponds to the numerical range of the symbol λ.

We apply the Gearhart–Prüss theorem to our operator $\mathcal{A}_D$, and our main theorem is the following.

Theorem 14.20

$$\widehat{\omega}_0(\mathcal{A}_D) = -\mathrm{Re}\,\lambda_1. \tag{14.5.13}$$

Using the eigenfunction associated with the leftmost eigenvalue λ_1, it is easy to see that

$$\|\exp -t\mathcal{A}^D\| \geq e^{-(\mathrm{Re}\,\lambda_1)t}. \tag{14.5.14}$$

Hence we immediately have

$$0 \geq \widehat{\omega}_0(\mathcal{A}^D) \geq -\mathrm{Re}\,\lambda_1. \tag{14.5.15}$$

To prove that $-\mathrm{Re}\,\lambda_1 \geq \widehat{\omega}_0(\mathcal{A}^D)$, it is enough to show the following lemma.

Lemma 14.21 *For any $\alpha < \operatorname{Re}\lambda_1$, there exists a constant C such that for all λ s.t. $\operatorname{Re}\lambda \leq \alpha$,*

$$\|(A^D - \lambda)^{-1}\| \leq C. \tag{14.5.16}$$

Proof This is clear by accretivity for $\alpha < 0$. For $\alpha > 0$, it is then enough to prove the lemma for $\operatorname{Re}\lambda \in [-\alpha, +\alpha]$. We know that λ is not in the spectrum. Hence the problem is just a check on the resolvent as $|\operatorname{Im}\lambda| \to +\infty$. The case where $\operatorname{Im}\lambda < 0$ has already been considered. Hence it remains to check the norm of the resolvent as $\operatorname{Im}\lambda \to +\infty$ when $\operatorname{Re}\lambda \in [-\alpha, +\alpha]$.

The proof is actually a semiclassical result! The main idea is that when $\operatorname{Im}\lambda$ tends to $+\infty$, we have to invert the operator

$$D_x^2 + i(x - \operatorname{Im}\lambda) - \operatorname{Re}\lambda.$$

If we consider the Dirichlet realization in the interval $S_\lambda :=]0, (\operatorname{Im}\lambda)/2[$ of $D_x^2 + i(x - \operatorname{Im}\lambda) - \operatorname{Re}\lambda$, it is easy to see that the operator is invertible by considering the imaginary part of this operator. Far from the boundary, we can use the resolvent of the problem on the line, for which we have a uniform control of the norm for $\operatorname{Re}\lambda \in [-\alpha, +\alpha]$.

More precisely, the main idea of the proof is to approximate $(A_D - \lambda)^{-1}$ by a sum of two different operators, one of which is a good approximation when applied to functions supported near the boundary, while the other takes care of functions whose support lies far away from the boundary. We denote the resolvent of the first operator by $R_1(\lambda)$ in $\mathcal{L}(L^2(S_\lambda, \mathbb{C}))$ and observe also that $R_1(\lambda) \in \mathcal{L}(L^2(S_\lambda, \mathbb{C}), H_0^1(S_\lambda, \mathbb{C}))$. We easily obtain the result that

$$\|R_1(\lambda)\| \leq \frac{2}{\operatorname{Im}\lambda}. \tag{14.5.17}$$

Furthermore, for $u = R_1 f$, we have

$$\|D_x R_1(\lambda) f\|^2 = \|D_x u\|^2 \leq \|(A_D - \lambda)u\|\|u\| + \operatorname{Re}\lambda\|u\|^2$$

$$\leq \|f\|\|R_1(\lambda)f\| + |\alpha|\|R_1(\lambda)f\|^2 \leq \left(\frac{2}{|\operatorname{Im}\lambda|} + \frac{4|\alpha|}{|\operatorname{Im}\lambda|^2}\right)\|f\|^2.$$

Hence there exists $C_0(\alpha)$ such that for $\operatorname{Im}\lambda \geq 1$ and $\operatorname{Re}\lambda \in [-\alpha, +\alpha]$,

$$\|D_x R_1(\lambda)\| \leq C_0(\alpha)|\operatorname{Im}\lambda|^{-1/2}. \tag{14.5.18}$$

Far from the boundary, we attempt to approximate the resolvent of the Airy operator $\overline{A_0}$ on the line considered in Section 14.3, neglecting the boundary effect. We denote this resolvent by $R_2(\lambda)$. Recall from (14.3.5) that the norm

$\|R_2(\lambda)\|$ is independent of Im λ. Since $R_2(\lambda)$ is an entire function in λ, we can easily obtain a uniform bound on $\|R_2(\lambda)\|$ for Re $\lambda \in [-\alpha, +\alpha]$. Hence,

$$\|R_2(\lambda)\| \leq C_1(\alpha). \tag{14.5.19}$$

In a similar manner to the derivation of (14.5.18) (see (14.3.1)), we can then show

$$\|D_x R_2(\lambda)\| \leq C(\alpha). \tag{14.5.20}$$

We now use a partition of unity in the variable x in order to construct an approximate inverse $R^{\mathrm{app}}(\lambda)$ for $\mathcal{A}_D - \lambda$. We shall prove that the difference between the approximation and the exact resolvent is well controlled as Im $\lambda \to +\infty$.

We define a pair (ϕ, ψ) of cutoff functions in $C^\infty(\mathbb{R}_+, [0, 1])$, satisfying

$$\phi(t) = 0 \text{ on } \left[0, \frac{1}{4}\right], \quad \phi = 1 \text{ on } \left[\frac{1}{2}, +\infty\right[,$$
$$\psi(t) = 1 \text{ on } \left[0, \frac{1}{4}\right], \quad \psi = 0 \text{ on } \left[\frac{1}{2}, +\infty\right[,$$
$$\phi(t)^2 + \psi(t)^2 \equiv 1 \text{ on } \mathbb{R}_+,$$

and then set

$$\phi_\lambda(x) = \phi\left(\frac{x}{\mathrm{Im}\,\lambda}\right), \quad \psi_\lambda(x) = \psi\left(\frac{x}{\mathrm{Im}\,\lambda}\right).$$

The approximate inverse $R^{\mathrm{app}}(\lambda)$ is then constructed as

$$R^{\mathrm{app}}(\lambda) = \psi_\lambda R_1(\lambda)\psi_\lambda + \phi_\lambda R_2(\lambda)\phi_\lambda, \tag{14.5.21}$$

where ϕ_λ and ψ_λ denote the operators of multiplication by the functions ϕ_λ and ψ_λ. Note that ψ_λ maps $L^2(\mathbb{R}_+)$ into $L^2(S_\lambda)$. In addition,

$$\psi_\lambda : H^2(S_\lambda) \cap H_0^1(S_\lambda) \to D(\mathcal{A}_D),$$
$$\phi_\lambda : D(\overline{\mathcal{A}_0}) \to D(\mathcal{A}_D).$$

From (14.5.17) and (14.5.19) we obtain, for sufficiently large Im λ,

$$\|R^{\mathrm{app}}(\lambda)\| \leq C_3(\alpha). \tag{14.5.22}$$

Note that

$$|\phi'_\lambda(x)| + |\psi'_\lambda(x)| \le \frac{C}{|\operatorname{Im}\lambda|}, \quad |\phi''_\lambda(x)| + |\psi''_\lambda(x)| \le \frac{C}{|\operatorname{Im}\lambda|^2}. \quad (14.5.23)$$

Next, we apply $\mathcal{A}_D - \lambda$ to R^{app} to obtain the result that

$$(\mathcal{A}_D - \lambda)R^{\mathrm{app}}(\lambda) = I + [\mathcal{A}_D, \psi_\lambda]R_1(\lambda)\psi_\lambda + [\mathcal{A}_D, \phi_\lambda]R_2(\lambda)\phi_\lambda, \quad (14.5.24)$$

where I is the identity operator, and

$$\begin{aligned}
[\mathcal{A}_D, \phi_\lambda] &:= \mathcal{A}_D\phi_\lambda - \phi_\lambda\mathcal{A}_D \\
&= [D_x^2, \phi_\lambda] \\
&= -\frac{2i}{\operatorname{Im}\lambda}\phi'\left(\frac{x}{\operatorname{Im}\lambda}\right)D_x - \frac{1}{(\operatorname{Im}\lambda)^2}\phi''\left(\frac{x}{\operatorname{Im}\lambda}\right).
\end{aligned} \quad (14.5.25)$$

A similar relation holds for $[\mathcal{A}_D, \psi_\lambda]$. Here, we have used (14.5.21) and the fact that

$$(\mathcal{A}_D-\lambda)R_1(\lambda)\psi_\lambda u = \psi_\lambda u, \quad (\mathcal{A}_D-\lambda)R_2(\lambda)\phi_\lambda u = \phi_\lambda u, \quad \forall u \in L^2(\mathbb{R}_+, \mathbb{C}).$$

Using (14.5.17), (14.5.18), (14.5.20), and (14.5.25), we then easily obtain, for sufficiently large $\operatorname{Im}\lambda$,

$$\|[\mathcal{A}_D, \psi_\lambda]R_1(\lambda)\| + \|[\mathcal{A}_D, \phi_\lambda]R_2(\lambda)\| \le \frac{C_4(\alpha)}{|\operatorname{Im}\lambda|}. \quad (14.5.26)$$

Hence, if $|\operatorname{Im}\lambda|$ is large enough, then $I + [\mathcal{A}_D, \psi_\lambda]R_1(\lambda)\psi_\lambda + [\mathcal{A}_D, \phi_\lambda]R_2(\lambda)\phi_\lambda$ is invertible in $\mathcal{L}(L^2(\mathbb{R}_+, \mathbb{C}))$, and

$$\left\|\left(I + [\mathcal{A}_D, \psi_\lambda]R_1(\lambda)\psi_\lambda + [\mathcal{A}_D, \phi_\lambda]R_2(\lambda)\phi_\lambda\right)^{-1}\right\| \le C_5(\alpha). \quad (14.5.27)$$

Finally, since

$$(\mathcal{A}_D - \lambda)^{-1} = R^{\mathrm{app}}(\lambda) \circ (I + [\mathcal{A}_D, \psi_\lambda]R_1(\lambda)\psi_\lambda + [\mathcal{A}_D, \phi_\lambda]R_2(\lambda)\phi_\lambda)^{-1},$$

we have the result that

$$\|(\mathcal{A}_D-\lambda)^{-1}\| \le \|R^{\mathrm{app}}(\lambda)\|\|\left(I+[\mathcal{A}_D, \psi_\lambda]R_1(\lambda)\psi_\lambda+[\mathcal{A}_D, \phi_\lambda]R_2(\lambda)\phi_\lambda\right)^{-1}\|.$$

Using (14.5.22) and (14.5.27), we find that (14.5.16) is true. $\qquad\square$

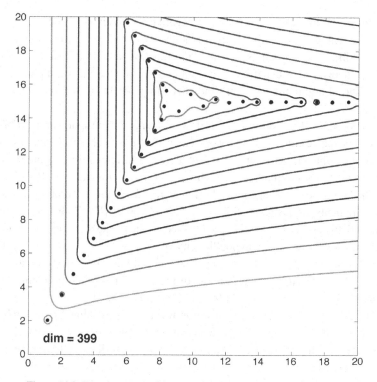

Figure 14.2 Pseudospectra of the complex Airy operator on the half-line.

The pseudospectra of the complex Airy operator on the half-line are shown in Figure 14.2.

14.6 Notes

The example treated in Section 14.1 was first presented in the book [TrEm]. The treatment in Section 14.2 was inspired by [Mart]. Via a Fourier transform, we recover the complex Airy operator, which was treated in Section 14.3. The Laplace method was invented by Pierre-Simon Laplace in 1774 and can be found in [Di]. A complete proof of Proposition 14.11 was given by W. Bordeaux-Montrieux [BM] (see also [Mart]). To the author's knowledge, the first mathematical analysis of the complex Airy operator appeared in an analysis of the Stark effect by I. Herbst [Her].

There have been many contributions to the content of Section 14.4, including publications by Aslayan and Davies [AsDa], Davies [Dav2, Dav3, Dav4], Boulton [Bou], Pravda-Starov [PS], Trefethen [Tr2], and Zworski [Zw], and references therein. Our goal here was to give a pedestrian proof. For the

Combes–Thomas argument [CT], readers are referred also to Reed and Simon [RS-IV]. Another possibility is to use the theory of ordinary differential equations [Sib]. The properties of the Airy function are described in [AbSt]. Kato's theory is presented in many standard books (see, for example, [Ka]). This kind of argument has been generalized to problems in superconductivity [AHP1, AHP2].

The proof of Proposition 14.13 follows from results presented in [PS, DSZ]. Note that the results in [DSZ] were given in the semiclassical limit for the spectral parameter in a compact set, but there is a simple scaling argument that allows one to pass to the limit of high frequency; see, for example, [Sj3, Sj2].

The estimate in (14.4.5) can be recovered from the general semiclassical statements of Dencker, Sjöstrand, and Zworski [DSZ]. It can also be obtained by using nilpotent Lie group techniques [HelNo] and the characterization of the maximal hypoellipticity of $-\partial_x^2 + (x^2 \partial_y + x \partial_z + \partial_s)$. The complex harmonic oscillator also appears naturally as a first approximation in the asymptotic (semiclassical) regime of the Orr–Sommerfeld problem in the form $-h^2(d^2/dx^2) + i(x^2 - 1)$ (see [TrEm, LanTr, DW]).

For Section 14.5, the main motivation was a paper by Almog [Alm], which was continued in [He7, AHP2]. Proposition 14.19 is proven in [Alm] (see also [Hen]). The zeros of the Airy function are described in [AbSt, section 10.4]. The pseudospectra of the complex Airy operator are illustrated in Figure 14.2. This figure is due to W. Bordeaux-Montrieux, who used the software package EigTool[4] to generate it. The figure shows the level curves $\mathcal{C}(\epsilon)$ of the norm of the resolvent $\|(A - z)^{-1}\| = 1/\epsilon$ corresponding to the boundary of the ϵ-pseudospectra. A. Jensen informed the author in June 2011 that this model was one of his favorite toy models for presenting the notion of pseudospectra.

14.7 Exercises

Exercise 14.1 (An example of PT-symmetry.) Let $J \in \mathbb{R}$. Show that a Dirichlet realization $\mathcal{A}_J$ in $]-1, +1[$ with domain $H^2(]-1, +1[) \cap H_0^1(]-1, +1[)$ can be associated with the complex Airy operator $D_x^2 + iJx$.

Recall what the spectrum for $J = 0$ is. Show that if (u, λ) is an eigenpair for $\mathcal{A}$, then $(Ku, \bar{\lambda})$ is an eigenpair for $\mathcal{A}_J$, where

$$Ku(x) = \bar{u}(-x).$$

4   See www-pnp.physics.ox.ac.uk/~stokes/courses/scicomp/eigtool/html/eigtool/
   documentation/menus/airy_demo.html and www.comlab.ox.ac.uk/pseudospectra/eigtool/.

Assume that the eigenvalues of $\mathcal{A}_J$ depend continuously on J. Show that for $|J|$ small enough, the eigenvalue with smallest real part is simple and real.[5]

Exercise 14.2 Analyze the Airy operators (real and complex) on $\mathbb{R}^+$ with a Neumann condition at 0.

Exercise 14.3 For $\theta \in]-\pi/2, +\pi/2[$, analyze the complex harmonic oscillator $-d^2/dx^2 + e^{i\theta}x^2$ on the line.

---

[5] This exercise was inspired by a talk given by P. Sternberg [RSZ] in Aarhus in June 2011. In "PT-symmetry," the letter P stands for "parity," i.e., the transformation $x \mapsto -x$, and the letter T stands for "time reversal," i.e., complex conjugation.

15

Applications in kinetic theory: the Fokker–Planck operator

The Kramers–Fokker–Planck operator arises from a simplification of some models that appear in kinetic theory. We intend to analyze the main spectral properties of this operator in this chapter. In this analysis, we shall encounter all of the concepts introduced in Chapter 13.

15.1 Definition and main results

Let V be a C^∞ potential on $\mathbb{R}^m$. We consider the operator K defined on $C_0^\infty(\mathbb{R}^{2m})$ by

$$K := -\Delta_v + \frac{1}{4}|v|^2 - \frac{m}{2} + X_0, \qquad (15.1.1)$$

where

$$X_0 := -\nabla V(x) \cdot \partial_v + v \cdot \partial_x. \qquad (15.1.2)$$

K is considered as an unbounded operator on the Hilbert space $\mathcal{H} = L^2(\mathbb{R}^{2m})$. The simplest model of this type is on $\mathbb{R}^2$ ($m = 1$) when we consider the quadratic potential $V(x) = (\tilde{\omega}_0^2/2)x^2$ (with $\tilde{\omega}_0 \neq 0$), for which rather explicit computations can be done:

$$-\frac{\partial^2}{\partial v^2} + \frac{1}{4}v^2 - \frac{1}{2} - \tilde{\omega}_0^2 x \partial_v + v \partial_x. \qquad (15.1.3)$$

These computations will be presented in Section 15.5.

In the general case, X_0 is the vector field generating the Hamiltonian flow associated with the Hamiltonian, $\mathbb{R}^m \times \mathbb{R}^m \ni (x, v) \mapsto \frac{1}{2}|v|^2 + V(x)$. Since K is accretive, it is closable. We denote its closure by $\overline{K}$, which we now call the Fokker–Planck operator. The first result is the following theorem.

Theorem 15.1 *For any $V \in C^{\infty}(\mathbb{R}^m)$, the associated Fokker–Planck operator is maximally accretive. Moreover, K^* is also maximally accretive and is the closure of the operator*

$$K_- := -\Delta_v + \frac{1}{4}|v|^2 - \frac{m}{2} - X_0, \qquad (15.1.4)$$

defined on $C_0^{\infty}(\mathbb{R}^{2m})$.

The second result is the following theorem.

Theorem 15.2 *Assume that, for some $\rho_0 > \frac{1}{3}$ and for $|\alpha| = 2$, there exists $C_\alpha > 0$ such that*

$$|D_x^\alpha V(x)| \leq C_\alpha (1 + |\nabla V(x)|^2)^{(1-\rho_0)/2} \qquad (15.1.5)$$

and

$$|\nabla V(x)| \to +\infty \quad as \ |x| \to +\infty. \qquad (15.1.6)$$

Then the Fokker–Planck operator $\overline{K}$ has a compact resolvent and there exists a constant $C > 0$ such that for all $v \in \mathbb{R}$,

$$|v|^{2/3} \, ||u||^2 + || \, |\nabla V(x)|^{2/3} \, u \, ||^2 \leq C \left(||(\overline{K} - iv)u||^2 + ||u||^2 \right),$$
$$\forall u \in C_0^{\infty}(\mathbb{R}^n). \qquad (15.1.7)$$

Corollary 15.3 *Uniformly for μ in a compact interval, there exist v_0 and C such that $\mu + iv$ is not in the spectrum of $\overline{K}$ and*

$$||(\overline{K} - \mu - iv)^{-1}|| \leq C \, |v|^{-1/3}, \ \forall v \ s.t. \ |v| \geq v_0. \qquad (15.1.8)$$

We shall show the following result as one of the main applications.

Theorem 15.4 *Under the assumptions of Theorem 15.2 and assuming that $e^{-V} \in L^1(\mathbb{R}^m)$, there exist $\alpha > 0$ and $\hat{C} > 0$ such that*

$$\forall u \in L^2(\mathbb{R}^{2m}), \quad \left\| e^{-t\overline{K}} u - \Pi_0 u \right\|_{L^2} \leq \hat{C} \, e^{-\alpha t} \, ||u||, \qquad (15.1.9)$$

where Π_0 is the projector defined for $u \in L^2(\mathbb{R}^{2m})$ by

$$(\Pi_0 u)(x, v) = \Phi(x, v) \left(\int \Phi(x, v) \, u(x, v) \, dx \, dv \right) \Big/ \left(\int \Phi(x, v)^2 \, dx \, dv \right),$$
$$(15.1.10)$$

with

$$\Phi(x, v) := \exp -\frac{v^2}{4} \exp -\frac{V(x)}{2}. \tag{15.1.11}$$

In addition, in Proposition 15.6, we shall discuss the set of α's such that (15.1.9) is satisfied.

15.2 Proof of Theorem 15.1: accretivity

The idea is to adapt the proof that a semibounded Schrödinger operator with a regular potential is essentially self-adjoint on $L^2(\mathbb{R}^n)$ (Theorem 9.15).

We apply the abstract criterion given by Theorem 13.14, taking $\mathcal{H} = L^2(\mathbb{R}^{2m})$ and $A = K$. Since the operators are real, that is, they respect real functions, we can consider functions that are everywhere real. The accretivity on $C_0^\infty(\mathbb{R}^n)$ is clear, using the positivity of the harmonic oscillator $-\Delta_v + \frac{1}{4}|v|^2 - m/2$. We can then consider the closure $\overline{K}$. Replacing K by $T := K + (m/2 + 1)I$, we have to show that its range is dense.

Let $f \in L^2(\mathbb{R}^n)$, with $n = 2m$, such that

$$\langle f, Tu \rangle_{\mathcal{H}} = 0, \ \forall u \in C_0^\infty(\mathbb{R}^n). \tag{15.2.1}$$

We have to show that $f = 0$. Because K is real, we can assume that f is real.

We first observe that (15.2.1) implies that

$$\left(-\Delta_v + \frac{1}{4}|v|^2 + 1 - X_0 \right) f = 0, \ \text{in } \mathcal{D}'(\mathbb{R}^n).$$

The standard hypoellipticity theorem for Hörmander operators of type 2 implies that $f \in C^\infty(\mathbb{R}^n)$. These Hörmander operators are of the form $P = -\sum_{j=1}^k X_j^2 + X_0 + a(x)$, where the X_j are real vector fields. If the X_j together with the brackets $[X_\ell, X_{\ell'}]$ span the whole tangent space at each point x, then it can be shown that the corresponding operator is hypoelliptic.[1] In our application, we have $k = 1$, $X_1 = \partial/\partial v$. The bracket $[X_1, X_0]$ gives $\partial/\partial x$ and the result applies.

We now introduce a family of cutoff functions $\zeta_k := \zeta_{k_1,k_2}$ by

$$\zeta_{k_1,k_2}(x, v) := \zeta(x/k_1)\zeta(v/k_2), \ \forall k \in \mathbb{N}^2, \tag{15.2.2}$$

where ζ is a C^∞ function satisfying $0 \le \zeta \le 1$, $\zeta = 1$ on $B(0, 1)$, and supp $\zeta \subset B(0, 2)$. For any $u \in C_0^\infty$, we have the identity

---

[1] P is said to be hypoelliptic if, for any $u \in \mathcal{D}'$ and any open set ω, $Pu \in C^\infty(\omega)$ implies that $u \in C^\infty(\omega)$.

$$\int \nabla_v(\zeta_k f) \cdot \nabla_v(\zeta_k u) \, dx \, dv + \int \zeta_k(x, v)^2 \left(\frac{v^2}{4} + 1 \right) u(x, v) \, f(x, v) \, dx \, dv$$

$$+ \int f(x, v)(X_0(\zeta_k^2 u))(x, v) \, dx \, dv$$

$$= \int |(\nabla_v \zeta_k)(x, v)|^2 u(x, v) f(x, v) \, dx \, dv$$

$$+ \sum_{i=1}^{m} \int \left(f(\partial_{v_i} u) - u(\partial_{v_i} f) \right)(x, v)\zeta_k(x, v)(\partial_{v_i} \zeta_k)(x, v) \, dx \, dv$$

$$+ \langle f(x, v), \, T\zeta_k^2 u \rangle. \tag{15.2.3}$$

When f satisfies (15.2.1), we obtain

$$\int_{\mathbb{R}^n} \nabla_v(\zeta_k f) \cdot \nabla_v(\zeta_k u) \, dx \, dv + \int \zeta_k^2 \left(\frac{|v|^2}{4} + 1 \right) u(x, v) \, f(x, v) \, dx \, dv$$

$$+ \int f(x, v)(X_0(\zeta_k^2 u))(x, v) \, dx \, dv$$

$$= \int |(\nabla_v \zeta_k)(x)|^2 u(x) f(x, v) \, dx \, dv \tag{15.2.4}$$

$$+ \sum_{i=1}^{m} \int \left(f(\partial_{v_i} u) - u(\partial_{v_i} f) \right)(x, v)\zeta_k(x, v)(\partial_{v_i} \zeta_k)(x, v) \, dx \, dv,$$

for all $u \in C^\infty(\mathbb{R}^n)$. In particular, we can take $u = f$. We obtain

$$\langle \nabla_v(\zeta_k f), \, \nabla_v(\zeta_k f) \rangle + \int \zeta_k^2 \left(\frac{|v|^2}{4} + 1 \right) |f(x, v)|^2 \, dx \, dv$$

$$+ \int f(x, v)(X_0(\zeta_k^2 f))(x, v) \, dx \, dv$$

$$= \int |\nabla_v \zeta_k|^2 |f(x, v)|^2 \, dx \, dv. \tag{15.2.5}$$

With an additional integration by parts, we obtain

$$\langle \nabla_v(\zeta_k f), \, \nabla_v(\zeta_k f) \rangle + \int \zeta_k^2 \left(\frac{|v|^2}{4} + 1 \right) |f(x, v)|^2 \, dx \, dv$$

$$+ \int \zeta_k f(x, v)^2 (X_0 \zeta_k)(x, v) \, dx \, dv$$

$$= \int |\nabla_v \zeta_k|^2 |f(x, v)|^2 \, dx \, dv. \tag{15.2.6}$$

This leads to the existence of a constant C such that, for all k,

$$||\zeta_k f||^2 + \frac{1}{4}||\zeta_k vf||^2$$
$$\leq \frac{C}{k_2^2}||f||^2 + \frac{C}{k_1}||v\zeta_k f|| \, ||f|| + \frac{C}{k_2}||\nabla V(x)\zeta_k f|| \, ||f||. \quad (15.2.7)$$

For a new constant C, we therefore have

$$||\zeta_k f||^2 + \frac{1}{8}||\zeta_k v f||^2 \leq C\left(\frac{1}{k_2^2} + \frac{1}{k_1^2}\right)||f||^2 + \frac{C(k_1)}{k_2}||\zeta_k f|| \, ||f||,$$
$$(15.2.8)$$

where

$$C(k_1) = \sup_{|x|\leq 2k_1} |\nabla_x V(x)|.$$

This implies

$$||\zeta_k f||^2 \leq C\left(\frac{\tilde{C}(k_1)}{k_2^2} + \frac{1}{k_1^2}\right)||f||^2. \quad (15.2.9)$$

This leads finally to $f = 0$. For example, we can first take the limit $k_2 \to +\infty$, which leads to

$$\left\|\zeta\left(\frac{x}{k_1}\right)f\right\|^2 \leq \frac{C}{k_1^2}||f||^2,$$

and then take the limit $k_1 \to +\infty$. $\qquad \square$

15.3 Analysis of toy models

15.3.1 A semiclassical toy model

We look first for an α-independent control of $P_{\alpha,h}^{-1}$, with

$$P_{\alpha,h} = -h^2\frac{d^2}{dt^2} + \frac{1}{4}t^2 + i(t - \alpha). \quad (15.3.1)$$

The goal is to prove the existence of positive constants C and h_0 such that for all $u \in C_0^\infty(\mathbb{R})$, $h \in \,]0, h_0]$, and $\alpha \in \mathbb{R}$,

$$||P_{\alpha,h}u|| \geq \frac{1}{C}h^{2/3}||u||. \quad (15.3.2)$$

First, we observe that

$$\operatorname{Re}\langle P_{\alpha,h}u,\, u\rangle = h^2\,\|u'\|^2 + \frac{1}{4}\,\|tu\|^2. \qquad (15.3.3)$$

Then we use a "multiplier" method and consider

$$\operatorname{Im}\left\langle P_{\alpha,h}u,\, \frac{(t-\alpha)}{\sqrt{h^{4/3}+(t-\alpha)^2}}u\right\rangle = \|(t-\alpha)(h^{4/3}+(t-\alpha)^2)^{-1/4}u\|^2$$
$$+ r_{\alpha,h}(u), \qquad (15.3.4)$$

with

$$r_{\alpha,h}(u) := -h^2\operatorname{Im}\left\langle\frac{d^2}{dt^2}u,\, (t-\alpha)(h^{4/3}+(t-\alpha)^2)^{-1/2}u\right\rangle. \qquad (15.3.5)$$

We rewrite this as

$$r_{\alpha,h}(u) = h^2\operatorname{Im}\left\langle\frac{d}{dt}u,\, \left[\frac{d}{dt},\, (t-\alpha)(h^{4/3}+(t-\alpha)^2)^{-1/2}\right]u\right\rangle,$$

which is estimated by

$$|r_{\alpha,h}| \le 2h^{4/3}\,\|u'\|\,\|u\|.$$

We obtain

$$2\|P_{\alpha,h}u\|\,\|u\| + 2h^{4/3}\|u'\|\,\|u\| \ge h^2\|u'\|^2 + \|(t-\alpha)(h^{4/3}+(t-\alpha)^2)^{-1/4}u\|^2.$$

We now consider the operator $Q_{\alpha,h}$, where

$$Q_{\alpha,h} := -h^2\frac{d^2}{dt^2} + (t-\alpha)^2(h^{4/3}+(t-\alpha)^2)^{-1/2}.$$

By a simple change of variables $t - \alpha = h^{4/3}t'$, we obtain the result that the lowest eigenvalue of this operator is $h^{2/3}\mu_1$, where μ_1 is the lowest eigenvalue of $Q_{0,1}$ ($\mu_1 > 0$). We obtain

$$2\|P_{\alpha,h}u\| + 2h^{4/3}\|u'\| \ge \mu_1 h^{2/3}\|u\|. \qquad (15.3.6)$$

Finally, we observe that for any $C_1 > 0$,

$$h^{4/3}\,\|u'\| \le h^{1/3}\,\|P_{\alpha,h}u\|^{1/2}\,\|u\|^{1/2} \le C_1\,\|P_{\alpha,h}u\| + \frac{h^{2/3}}{C_1}\,\|u\|. \qquad (15.3.7)$$

Choosing C_1 large enough in (15.3.7) and using (15.3.6), this achieves the proof of the inequality (15.3.2).

In a similar way, we can also prove the optimal regularity result,

$$\left\|\left(-h^2\frac{d^2}{dt^2}+\frac{1}{4}t^2\right)u\right\|^2 + \|(t-\alpha)u\|^2 \leq \|P_{\alpha,h}u\|^2 + Ch^2\,\|u'\|\,\|u\|$$

$$\leq \hat{C}\|P_{\alpha,h}u\|^2,$$

for h small enough. The above inequalities can be combined with

$$\sum_{k+\ell\leq 2} h^{2-(k+\ell)}\left\|t^k\left(h\frac{d}{dt}\right)^\ell u\right\|^2 \leq C'\left\|\left(h^2\frac{d^2}{dt^2}+\frac{1}{4}t^2\right)u\right\|^2. \tag{15.3.8}$$

When h is small enough, we also obtain the estimate (15.3.2) for $\widehat{P}_{\alpha,h} = P_{\alpha,h} - h/2$.

Lemma 15.5 *There exist h_0, C, and C' such that for all $h \in\,]0, h_0]$ and all $\alpha \in \mathbb{R}$,*

$$\|\widehat{P}_{\alpha,h}^{-1}\| \leq C\,h^{-2/3} \tag{15.3.9}$$

and

$$h^{4/3}\,\|u\|^2 + \sum_{k+\ell\leq 2} h^{2-(k+\ell)}\left\|t^k\left(h\frac{d}{dt}\right)^\ell u\right\|^2 \leq C'\,\|\widehat{P}_{\alpha,h}u\|^2, \tag{15.3.10}$$

for all $u \in C_0^\infty(\mathbb{R})$.

15.3.2 Reinterpretation after scaling

We now consider

$$\widehat{F}_{\rho,v} := -\frac{d^2}{dv^2} + \frac{1}{4}v^2 - \frac{1}{2} + i(\rho v - v). \tag{15.3.11}$$

Introducing $v = \rho t$, $\alpha = v/\rho^2$, and $h = \rho^{-2}$, we see that

$$\widehat{F}_{\rho,v} = \rho^2 P_{\alpha,h},$$

and Lemma 15.5 gives the existence of $C > 0$ such that

$$\left\|(\widehat{F}_{\rho,v})^{-1}\right\| \leq C\rho^{-2/3}, \ \forall v,\, \forall \rho > C. \tag{15.3.12}$$

This is not exactly what we want, because we need a check for $|v|$ large, independently of ρ. But for $|v| \geq (1/C_2)\rho^2$, we have

$$\left\|\left(\widehat{F}_{\rho,v}\right)^{-1}\right\| \leq C_3 \, |v|^{-1}. \tag{15.3.13}$$

Hence, altogether, we obtain

$$\left\|\left(\widehat{F}_{\rho,v}\right)^{-1}\right\| \leq C_4(|v| + \rho^2)^{-1/3}, \tag{15.3.14}$$

for $\rho^2 + |v| \geq C$.

Using (15.3.8), we can also have a check of

$$\left\|t\left(\widehat{F}_{\rho,v}\right)^{-1}\right\|$$

and

$$\left\|\frac{d}{dt}\left(\widehat{F}_{\rho,v}\right)^{-1}\right\|$$

in $\mathcal{L}(L^2(\mathbb{R}))$.

15.3.3 The model on $\mathbb{R}^2$

For a fixed $w \in \mathbb{R}$, we now consider the model on $\mathbb{R}^2$

$$F_w := -\partial_v^2 + \frac{1}{4}v^2 - \frac{1}{2} + (v\partial_x - w\partial_v). \tag{15.3.15}$$

To use the preceding results, we first take a partial Fourier transform in the variable x and obtain the family

$$F_{w,\xi} := -\frac{d^2}{dv^2} + \frac{1}{4}v^2 - \frac{1}{2} + \left(iv\xi - w\frac{d}{dv}\right). \tag{15.3.16}$$

Let us show that this operator is unitarily equivalent to

$$\widehat{F}_\rho := -\frac{d^2}{dv^2} + \frac{1}{4}v^2 - \frac{1}{2} + i\frac{\rho}{2}v, \tag{15.3.17}$$

with

$$\rho = \sqrt{4\xi^2 + w^2}. \tag{15.3.18}$$

We write $2\xi = \rho\hat{\xi}$ and $w = \rho\hat{\eta}$, and so $|\hat{\eta}|^2 + |\hat{\xi}|^2 = 1$. Then, with

$$X_1 = i\frac{v}{2}\hat{\xi} - \hat{\eta}\frac{d}{dv} \quad \text{and} \quad X_2 = i\frac{v}{2}\hat{\eta} + \hat{\xi}\frac{d}{dv},$$

we rewrite $F_{w,\xi}$ in the form

$$F_{w,\xi} = -X_1^2 - X_2^2 - \frac{1}{2} + \rho X_1.$$

By a finite succession of gauge transformations with quadratic phase, Fourier transforms, and dilations starting with $u \mapsto e^{i(\hat{\xi}/4\hat{\eta})v^2}u$, it is then easy to obtain the result.

From (15.3.14) we obtain, for $|v| \geq C$,

$$||(F_{w,\xi} - iv)^{-1}|| \leq C_4\,(|v| + 4\xi^2 + w^2))^{-1/3}.$$

It is possible to show the existence of a constant $C > 0$ such that for any $\rho = \sqrt{w^2 + 4\xi^2} \geq C$, $v \in \mathbb{R}$, the following maximal estimate holds:

$$||u||^2 + ||(F_{w,\xi} - iv)u||^2 \geq \frac{1}{C}\left((|v| + \rho^2)^{2/3}\,||u||^2 + \sum_{k+\ell\leq 2} ||D_v^k v^\ell u||^2\right),$$

$$\forall u \in \mathcal{S}(\mathbb{R}). \tag{15.3.19}$$

But the case where $\rho \leq C$ can be treated easily. Hence, taking $\xi = 0$ in the right-hand side of (15.3.19), and after a partial Fourier transform, we obtain

$$||(F_w - iv)^{-1}|| \leq C_4\,(|v| + w^2))^{-1/3}.$$

We therefore obtain the existence of a constant $C > 0$ such that for any $v \in \mathbb{R}$, the following maximal estimate is valid:

$$||u||^2 + ||F_w^v u||^2 \geq \frac{1}{C}\left((|v| + w^2)^{2/3})||u||^2 + \sum_{k+\ell\leq 2} ||D_v^k v^\ell u||^2\right),$$

$$\forall u \in C_0^\infty(\mathbb{R}^2), \tag{15.3.20}$$

with

$$F_w^v := F_w - iv.$$

Note that, by interpolation, this implies also that

$$||u||^2 + ||(F_w^\nu)u||^2 \geq \frac{1}{C}\left(|w|^{2/3}||\nu u||^2 + |w|^{2/3}||\partial_\nu u||^2\right), \quad \forall u \in C_0^\infty(\mathbb{R}^2).$$
$$(15.3.21)$$

15.4 Proof of Theorem 15.2: estimates of resolvents

The approach is the following. We first locally replace $\nabla V(x)$ by a constant vector w, prove global estimates for this local model F_w (actually $F_w - i\nu$), and then patch together the estimates, using a partition of unity and controlling the localization errors.

We now present this approach when $n = 1$ for simplicity (but this restriction is not important). We introduce a partition of unity ϕ_j in the variable x corresponding to a covering by intervals $I(x_j, r(x_j)) =]x_j - r(x_j), x_j + r(x_j)[$ (with the property of uniform finite intersection, that is, the existence of an integer n_0 such that any point is covered by fewer than n_0 intervals belonging to the covering family), where $r(x)$ has the expression

$$r(x) := \delta_0 \langle V'(x)\rangle^{-1/3}. \qquad (15.4.1)$$

Here, $\langle V'(x)\rangle$ is the Japanese bracket of V',

$$\langle V'(x)\rangle := \sqrt{1 + |V'(x)|^2},$$

and $\delta_0 \geq 1$ is an extra parameter that will be chosen later. So, the support of each ϕ_j is contained in $I(x_j, r(x_j))$, and we have

$$\sum_j \phi_j(x)^2 = 1. \qquad (15.4.2)$$

Moreover, there exists another partition of unity χ_j with support in $I(x_j, 2r(x_j))$, a constant C such that for all j and all δ_0,

$$\sum \chi_j(x)^2 \leq C,$$

and

$$|\phi_j'(x)| \leq \frac{C}{\delta_0}\langle V'(x_j)\rangle^{1/3}\chi_j(x).$$

Note that we can find C_{δ_0} such that for $|x_j| \geq C_{\delta_0}$, we have

$$\frac{1}{2} \leq \frac{|V'(x)|}{|V'(x_j)|} \leq \frac{3}{2}, \quad \forall x \in I(x_j, 2r(x_j)).$$

This introduces two types of errors:

- The first type is due to the comparison of $V'(x)$ and $V'(x_0)$ in the interval, leading to an error in $V''(x_0)r(x_0)$, which has to be small in comparison with the gain $|V'(x_0)|^{1/3}$. This leads to the assumption that $|V''(x_0)|\, r(x_0)$ will be controlled by $\langle V'(x_0)\rangle^{2/3}$.
- The second type of error is due to the partition of unity ϕ_j. The typical error to be considered here is $||\phi_j'(x)yf||^2$, which will be controlled if $r(x)^{-1}$ is controlled by $\langle V'(x)\rangle^{1/3}$. This explains our choice of radius and our assumptions about $V''(x)$.

Partition of unity Let us now give a detailed proof. We start, for $u \in C_0^\infty(\mathbb{R}^2)$, from

$$||(K - iv)u||^2 = \sum_j ||\phi_j(K - iv)u||^2 \tag{15.4.3}$$

$$= \sum ||(K - iv)\phi_j u||^2 - \sum ||[K, \phi_j]u||^2 \tag{15.4.4}$$

$$= \sum ||(K - iv)\phi_j u||^2 - \sum ||(X_0\phi_j)u||^2 \tag{15.4.5}$$

$$= \sum ||(K - iv)\phi_j u||^2 - \sum ||\phi_j'(x)vu||^2 \tag{15.4.6}$$

$$\geq \sum ||(K - iv)\phi_j u||^2 - \frac{C^2}{\delta_0^2} \sum ||\chi_j\langle V'(x_j)\rangle^{1/3}vu||^2. \tag{15.4.7}$$

Let us now write, on the support of ϕ_j,

$$K - iv = \left(K - F_{w_j}\right) + F_{w_j}^v,$$

with $w_j = V'(x_j)$. We can verify that

$$||(K - F_{w_j})\phi_j u||^2 = ||\phi_j(x)(V'(x) - V'(x_j))\, \partial_v u||^2 \tag{15.4.8}$$

$$\leq C\delta_0^2\langle V'(x_j)\rangle^{2-2\rho_0-2/3}||\phi_j \partial_v u||^2. \tag{15.4.9}$$

These errors are controlled by the main term. We note that we have, from (15.3.20) and (15.3.21),

$$||F_{w_j}^v\phi_j u||^2 \geq \frac{1}{C} ||\, |V'(x)|^{2/3}\phi_j u||^2 + \frac{1}{C} ||\, |V'(x)|^{1/3}\phi_j \partial_v u||^2 \tag{15.4.10}$$

$$+ \frac{1}{C} ||\, |V'(x)|^{1/3}\phi_j vu||^2 + \frac{1}{C} |v|^{2/3} ||\phi_j u||^2. \tag{15.4.11}$$

Finally, we observe that

$$\|(K - iv)\phi_j u\|^2 \geq \frac{1}{2}\|F_{w_j}^v \phi_j u\|^2 - \|(K - F_{w_j})\phi_j u\|^2.$$

By summing over j, we obtain the existence of a constant C such that, for any $\delta_0 \geq 1$ and for all $u \in C_0^\infty(\mathbb{R}^2)$,

$$\|u\|^2 + \|(K - iv)u\|^2$$
$$\geq \frac{1}{C}\| |V'(x)|^{2/3} u \|^2 + \frac{1}{C}\| |V'(x)|^{1/3} \partial_v u \|^2 + \frac{1}{C}\| |V'(x)|^{1/3} vu \|^2$$
$$+ \frac{1}{C}|v|^{2/3}\|u\|^2 - C\delta_0^2\| |V'(x)|^{2/3-\rho_0} \partial_v u \|^2 - C\frac{1}{\delta_0^2}\| |V'(x)|^{1/3} \partial_v u \|^2$$
$$- C\delta_0(\|vu\|^2 + \|u\|^2 + \|\partial_v u\|^2). \tag{15.4.12}$$

If we choose δ_0 large enough, we can achieve the proof by observing that $|\nabla V(x)|$ tends to $+\infty$ and that $\rho_0 > \frac{1}{3}$.

Compact resolvent We have already mentioned the hypoellipticity of K. The proof of this hypoellipticity shows actually that $D(\overline{K}) \subset H_{\text{loc}}^{2/3}$ (with continuous injection). It is also possible to give a direct proof starting from (15.3.19). The second point is to show the existence of a weight $\rho(x, v)$ tending to $+\infty$ as $|x| + |v| \to +\infty$ such that $D(\overline{K}) \subset L_\rho^2$ with continuous injection. Here we can take

$$\rho(x, v) = v^2 + |\nabla V(x)|^{4/3},$$

which tends to $+\infty$, by our assumption about ∇V. Then we can use Proposition 5.4 to show the compact injection of $D(\overline{K})$ into L^2. $\qquad\square$

15.5 Explicit computation of the spectrum of the quadratic Fokker–Planck operator

We consider the case where $m = 1$ and $V(x) = (\tilde{\omega}_0^2/2)x^2$. After a dilation in the variable x, the operator considered in (15.1.3) becomes

$$L = -\frac{d^2}{dv^2} + \frac{1}{4}v^2 - \frac{1}{2} - \tilde{\omega}_0(v\partial_x - x\partial_v). \tag{15.5.1}$$

As proven in the previous section, this operator has a compact resolvent; we wish to determine its spectrum.

With

$$b = \partial_v + \frac{1}{2}v, \quad a = \partial_x + \frac{1}{2}x,$$

this operator can also be written as

$$L = b^*b + \tilde{\omega}_0(b^*a - a^*b). \tag{15.5.2}$$

We introduce

$$\delta = \sqrt{1 - 4\tilde{\omega}_0^2}.$$

So, there are two main cases, according to the sign of $(1 - 4\omega_0^2)$, and a special case corresponding to $\tilde{\omega}_0 = \pm\frac{1}{2}$.

When $|\tilde{\omega}_0| > \frac{1}{2}$, the two eigenvalues are complex conjugates, i.e., $\lambda_2 = \bar{\lambda}_1$. A possible way to determine the spectrum (see Problem 16.11 for another approach) is to expand a function u in $L^2(\mathbb{R}^2)$ in the basis of the eigenfunctions of the harmonic oscillator, that is, in the basis of the usual Hermite functions $h_{k_1,k_2}(x_1, x_2) = h_{k_1}(x_1)h_{k_2}(x_2)$ of two variables (see (1.3.8)), and to observe that for a given N, the spaces V_N generated by the h_{k_1,k_2} with $k_1 + k_2 = N + 1$ ($k_1 = 1, \ldots, N$) are stable. We can choose the eigenfunctions such that

$$ah_{k_1,k_2} = (k_1 - 1)h_{k_1-1,k_2}, \quad a^*h_{k_1,k_2} = h_{k_1+1,k_2},$$
$$bh_{k_1,k_2} = (k_2 - 1)h_{k_1-1,k_2}, \quad b^*h_{k_1,k_2} = h_{k_1+1,k_2}.$$

We then have to analyze the restriction of the operator L to each V_N, which is an $N \times N$ matrix, whose eigenvalues can be explicitly computed.

The eigenvalue equation takes the following form, for $y = (y_1, \ldots, y_N)$ (y_n being the coordinate of the vector corresponding to $h_{n,N+1-n}$) and with the convention that $y_0 = y_{N+1} = 0$:

$$(n - 1)\tilde{\omega}_0 y_{n-1} + (n - \lambda)y_n - (N - n)\tilde{\omega}_0 y_{n+1} = 0, \tag{15.5.3}$$

for $n = 1, \ldots, N$. If $|\tilde{\omega}_0| \neq \frac{1}{2}$, the matrix of $L_{/V_N}$ is diagonalizable; its eigenvalues have multiplicity 1 and are given by

$$\lambda_{n_1,n_2} = \frac{N + 1}{2} + \frac{1}{2}\delta(n_1 - n_2), \tag{15.5.4}$$

with $n_1 + n_2 = N + 1$. Hence the spectrum is real if $0 < |\tilde{\omega}_0| < \frac{1}{2}$ and complex if $|\tilde{\omega}_0| > \frac{1}{2}$. In the case where $\tilde{\omega}_0 = \pm\frac{1}{2}$, the matrix has a unique eigenvalue $N/2$ (with algebraic multiplicity N), and one can only write a Jordan form.

It remains to verify that the family of eigenfunctions is total in $L^2(\mathbb{R}^2)$. But this was done for the h_{k_1} in the introduction, and the result then follows easily.

15.6 Proof of Theorem 15.4: growth of the semigroup

Note that when $\exp -V$ is in L^1, we have

$$\mathrm{Ker}\,(\overline{K}) = \mathbb{C}\Phi(x, v). \tag{15.6.1}$$

On the other hand, we always have

$$\mathrm{Ker}\,\overline{K} = \mathrm{Ker}\,K^* \tag{15.6.2}$$

and

$$\overline{K}\,\mathrm{Ker}\,\overline{K}^{\perp} \subset \mathrm{Ker}\,\overline{K}^{\perp}. \tag{15.6.3}$$

We have already proved that the Fokker–Planck operator $\overline{K}$ has a compact resolvent. We have only to check that $\overline{K}u = ivu$ with $v \in \mathbb{R}$ and u in the domain of $\overline{K}$ implies $v = 0$. The relation $\mathrm{Re}\,\langle \overline{K}u \,|\, u\rangle = 0$ implies $u(x, v) = \varphi(x)\,e^{-v^2/4}$ and $\overline{K}u = \sum_{i=1}^{m}\left((\partial_{x_i} + x_i/2)\varphi\right) v_i\,e^{-v^2/4}$, which is possible only if $(\partial_{x_i} + x_i/2)\varphi = 0$ and $v = 0$.

We can then apply the Gearhart–Prüss theorem (Theorem 13.26) to the restriction $\overline{K}^{\mathrm{res}}$ of $\overline{K}$ to $\mathrm{Ker}\,\overline{K}^{\perp}$. This restriction has no spectrum on the imaginary axis, and Corollary 15.3 gives a control on the resolvent for $|v|$ large. This implies, in particular, the theorem. $\qquad\Box$

We can actually improve the conclusion of the theorem. We observe that $\overline{K}^{\mathrm{res}}$ has no spectrum in the half-plane $\mathrm{Re}\,z < \mathrm{Re}\,\lambda_2$, where λ_2 denotes the nonzero eigenvalue of $\overline{K}$ with smallest real part. Using Corollary 15.3 for $\mu \in [-1, \mathrm{Re}\,\lambda_2 - \epsilon]$ (with $\epsilon > 0$) and the accretivity estimate for $\mu \le -1$, we obtain the following proposition.

Proposition 15.6 *If λ_2 denotes the nonzero eigenvalue of $\overline{K}$ with smallest real part, then for any $\alpha < \mathrm{Re}\,\lambda_2$, there exists a constant C_α such that (15.1.9) is satisfied.*

Remark 15.7 In (15.1.9), we can take

$$C_\alpha = 1 + 2\frac{|\alpha|}{r(\alpha)}, \tag{15.6.4}$$

with

$$\frac{1}{r(\alpha)} = \sup_{\mathrm{Re}\,z \le \alpha} \|(\overline{K} - z)^{-1}\|.$$

This is a consequence of Proposition 13.31, with $\hat{\omega} = 0$, $\omega = -\alpha$, and $A = -\overline{K}$. Moreover, we have, by (13.5.8),

$$\frac{1}{r(\alpha)} = \sup_{\nu \in \mathbb{R}} ||(\overline{K} - \alpha - i\nu)^{-1}||.$$

15.7 On a link between Fokker–Planck operators and Witten Laplacians

Witten Laplacians $-\Delta_V^w$ have already appeared (modulo conjugation) in Exercise 7.7. In this chapter, we deal with only the Witten Laplacian on functions, which is given for $V \in C^\infty(\mathbb{R}^m)$ by

$$-\Delta_V^w = -\Delta + \frac{1}{4}|\nabla V|^2 - \frac{1}{2}\Delta V. \tag{15.7.1}$$

If we observe that

$$\langle -\Delta_V^w u, u \rangle = \left\|\left(\nabla + \frac{1}{2}\nabla V\right)u\right\|^2 \geq 0, \tag{15.7.2}$$

it is clear that $-\Delta_V^w$ defined on $C_0^\infty(\mathbb{R}^m)$ is essentially self-adjoint, by Theorem 9.15, and from now on we shall call this extension simply the Witten Laplacian.

It is natural to want to determine the assumptions under which this operator has a compact resolvent. We do not pretend to achieve optimality in this book, but it is easy to see that under the assumptions

$$|\Delta V(x)| \leq C < \nabla V(x) >^{2-\epsilon_0} \tag{15.7.3}$$

and

$$|\nabla V(x)| \to +\infty, \quad \text{as } |x| \to +\infty,$$

$-\Delta_V^w$ has a compact resolvent. In this case, we denote the sequence of its eigenvalues by ω_j ($j \in \mathbb{N}^*$). Assuming that e^{-V} belongs to L^1, it is also easy to verify that 0 is an eigenvalue of $-\Delta_V^w$:

$$\omega_1 = 0.$$

We shall just give an illustrative example of some relations between the decay of the semigroup associated with $\overline{K}$ and ω_2.

If g_0 is chosen as

$$g_0 = \Psi(x)(2\pi)^{-m/4} \exp -\frac{v^2}{4},$$

where $\Psi(x)$ is a normalized eigenfunction of $-\Delta_V^w$ corresponding to the eigenvalue $\omega_2 > 0$, then, if we assume that for some $\alpha > 0$ there exists $C > 0$ such that

$$\|\exp -t\overline{K}g\| \le C \exp -\alpha t \|g\|, \tag{15.7.4}$$

for any g orthogonal to Φ, then we obtain the following inequality:

$$\begin{aligned}
\frac{d}{dt}\|\exp -t\overline{K}g_0\|^2 &= -2\operatorname{Re}\langle \overline{K} \exp -t\overline{K}g_0 \mid \exp -t\overline{K}g_0\rangle \\
&\ge -2\|\overline{K}g_0\| \, \|\exp -tK^* \exp -t\overline{K}g_0\| \\
&\ge -2\sqrt{m+2}\sqrt{\omega_2}\, C^2 \exp -2\alpha t.
\end{aligned} \tag{15.7.5}$$

Integrating between 0 and t leads to

$$1 - \|\exp -t\overline{K}g_0\|^2 \le \frac{1}{\alpha}\sqrt{m+2}\sqrt{\omega_2}\, C^2 (1 - \exp -2\alpha t).$$

Taking the limit $t \to +\infty$, this leads to

$$\alpha \le \sqrt{m+2}\sqrt{\omega_2}\, C^2. \tag{15.7.6}$$

We now have to implement the information given by the Gearhart–Prüss theorem, which relates the optimal values of α and $\operatorname{Re}\lambda_2$ to the accurate values, and thereby gives information about the constant C. Of course, this is only interesting when the constant C is controlled. One expects that in the semiclassical case the constant C will be controlled by some negative power of h. This means that in this case, if ω_2 can be shown to be exponentially small, then one can expect that the best α will necessarily have the same property.

More precisely, for any $\alpha < \operatorname{Re}\lambda_2$, we obtain the following proposition using (15.6.4).

Proposition 15.8 *Under the assumptions of Theorem 15.2, for any $\alpha < \operatorname{Re}\lambda_2$, we have*

$$\frac{\alpha}{(1 + 2|\alpha|/r(\alpha))^2} \le \sqrt{m+2}\sqrt{\omega_2}. \tag{15.7.7}$$

15.8 Notes

For this chapter, readers are referred to the publications of Risken [Ris], Hérau and Nier [HerN], Helffer and Nier [HelN], and Villani [Vi], and the references therein. Although there are no new results in this chapter, we have presented proofs here that avoid the use of microlocal analysis or the use of nilpotent techniques. But readers need to know that these elementary proofs have limits and sometimes hide what is going on. In continuation of [HerN] and the work of Eckmann, Pillet, and Rey-Bellet [EcPRB], Villani's method (sometimes called the hypocoercivity method on the suggestion of T. Gallay) consists in comparing under suitable conditions the spectral properties of $A^*A + B$, where B is antiself-adjoint, with the properties of $A^*A + C^*C$, where $C = [A, B]$. This applies to the case of the Fokker–Planck operator on $\mathbb{R}^2$ with $A = \partial_v - v/2$ and $B = v\partial_x - \partial_x V \partial_v$. We obtain $C = \partial_x - \frac{1}{2}\partial_x V$ and recognize $A^*A + C^*C$ in this case as the Witten Laplacian associated with the function $(x, v) \mapsto v^2/2 + V(x)$.

The partitions of unity appearing in Section 15.4 are presented in a more general context in Hörmander's book [Ho2, chapter 1, section 1.4], in relation to the notion of a slowly varying metric. The results on the hypoellipticity of (minus) a sum of squares of vector fields are due to Hörmander [Ho1].

The subject covered by this chapter is a very active research field. We should mention the publications by Hérau, Sjöstrand, and Stolk [HerSjSt], and Hérau, Hitrik, and Sjöstrand [HHS1,HHS2], where a very accurate semiclassical analysis was given. There are various conjectures about a link between the Witten Laplacians, for which there is a huge semiclassical literature, and the Fokker–Planck operator (see [HelN] for references). The last section of the chapter follows [HelN], with the difference that we have implemented what we know from the Gearhart–Prüss theorem. Finally, we should mention Wen Deng's Ph.D. thesis [DW], which includes a survey of contributions in recent years, including the paper by Gallagher, Gallay, and Nier [GGN], which also gives an analysis of the model $\widehat{F}_{\rho,\nu}$ that appears in our analysis but in the case $\nu = 0$. An explicit constant C in (15.7.4) has been discussed in [HerN], [HelN], and all the recent semiclassical literature mentioned above.

15.9 Exercises

Exercise 15.1 (From X. Pan, personal communication.) Show that the spectrum of the operator

$$\mathcal{P} = -\Delta_{x_1,x_2} - 2ix_2\partial_{x_1} + iJx_2 + x_2^2$$

in $\mathbb{R}^2$ is given, if $J \neq 0$, by

$$\sigma(\mathcal{P}) = \bigcup_{n} \left\{ (2n+1) + \frac{J^2}{4} + i\mathbb{R} \right\}.$$

Exercise 15.2 (See [Alm, Hen].) Analyze the spectrum of the Dirichlet realization of $-\Delta_{x,y} + i(x+y)$ in $\mathbb{R}^+ \times \mathbb{R}^+$. Show that it has a compact resolvent, and compute the spectrum using the structure of the operator with separate variables.

Hint: Use Section 14.5 and Proposition 14.19.

Exercise 15.3 (The Kolmogorov operator.) Consider in $\mathbb{R}^3$ the operator defined on $C_0^\infty(\mathbb{R}^3)$ by

$$P_0 := -\partial_x^2 + (\partial_y + x\partial_z).$$

Show that this operator is maximally accretive, and determine the spectrum of its closure.

16

Problems

Problem 16.1

Part I[1]

Let $\mathcal{H} := L^2(\mathbb{S}^1)$ be the space of the complex L^2 functions on the circle (which can be identified with $L^2(]0, 2\pi[)$). For $u \in \mathcal{H}$, we introduce

$$||u||_{\mathcal{H}}^2 = \frac{1}{2\pi} \int_0^{2\pi} |u(\theta)|^2 \, d\theta.$$

For $u \in \mathcal{H}$ and $n \in \mathbb{Z}$, we introduce

$$u_n = \frac{1}{2\pi} \int_0^{2\pi} u(\theta) \exp -in\theta \, d\theta,$$

and we recall that the operator, denoted by $\mathcal{F}$, which associates with u in $\mathcal{H}$ the sequence of its Fourier coefficients $(u_n)_{n \in \mathbb{Z}}$ is a unitary operator from $\mathcal{H}$ onto $\ell^2(\mathbb{Z})$. In particular,

$$||u||_{\mathcal{H}}^2 = \sum_{n \in \mathbb{Z}} |u_n|^2.$$

We consider on $\mathcal{H}$ the operator P defined by

$$(Pu)_n = \begin{cases} llu_n & \text{for } n \geq 0, \\ 0 & \text{for } n < 0. \end{cases}$$

(a) Show that P is in $\mathcal{L}(\mathcal{H})$ and satisfies $P = P^* = P^2$ and $||P|| = 1$.

---

[1] This problem was inspired by a course given by G. Lebeau at the Ecole Polytechnique.

(b) After verification that $\overline{u_{-n}} = (\bar{u})_n$ for all $n \in \mathbb{Z}$, show that if u is real-valued, then

$$||Pu||_{\mathcal{H}} \geq \frac{1}{\sqrt{2}}||u||_{\mathcal{H}}.$$

(c) Let $C^0(\mathbb{S}^1)$ be the space of the continuous functions on $\mathbb{S}^1$ (which can be identified with the (2π)-periodic continuous functions on $\mathbb{R}$). Show that $C^0(\mathbb{S}^1)$ has the natural structure of a Banach space, for a suitable norm.

(d) Let $a \in C^0(\mathbb{S}^1)$. Show that the application which associates with $a \in C^0$ the operator of multiplication by a, denoted by $M(a)$ (or, more simply, by a), is continuous from $C^0(\mathbb{S}^1)$ into $\mathcal{L}(\mathcal{H})$.

(e) Show briefly that $C^\infty(\mathbb{S}^1)$ is dense in C^0. Deduce (or show directly) that the space $C^{0,f}(S^1)$ of the trigonometric polynomials is dense in $C^0(S^1)$.

(f) Show that if $a \in C^\infty(\mathbb{S}^1)$ and $u \in \mathcal{H}$, then

$$(au)_n = \sum_{m \in \mathbb{Z}} a_{n-m} u_m.$$

(g) For $a \in C^0(\mathbb{S}^1)$, we define $T(a)$ by

$$T(a) = PM(a)P.$$

Show that $T(a) \in \mathcal{L}(\mathcal{H})$ and that if q is real, then $T(a)$ is self-adjoint.

(h) Show that if $a \in C^{0,f}(\mathbb{S}^1)$, the operator $M(a)P - PM(a)$ has finite rank.

(i) Show that if $a \in C^0(\mathbb{S}^1)$, the operator $M(a)P - PM(a)$ is compact from $\mathcal{H}$ into $\mathcal{H}$.

(j) Show that, for any a_1, a_2 in $C^0(\mathbb{S}^1)$, there exists a compact operator K such that

$$T(a_1)T(a_2) = T(a_1a_2) + K.$$

(k) Show that if $\lambda \notin a(\mathbb{S}^1) \cup \{0\}$, then $T(a)$ cannot admit a Weyl sequence at λ. (We suggest that you use (j) with $a_2 = (a - \lambda)$ and a suitable a_1.)

(l) From now on, we assume that a is real. Show that if $\lambda \in a(\mathbb{S}^1)$, then one can construct a sequence of real functions $u_n \in \mathcal{H}$ such that $||u_n|| = 1$, u_n tends weakly to 0, and $(M(a) - \lambda)u_n$ tends strongly to 0.

(m) Let $\lambda \in a(\mathbb{S}^1)$. Show that the image under P of the sequence (u_n) constructed in (l) is a Weyl sequence for $T(a)$ at λ. (We suggest that you use (b) and (i) here.)

(n) Conclude that $\sigma_{ess}(T(a)) = a(\mathbb{S}^1) \cup \{0\}$, where $\sigma_{ess}(T(a))$ denotes the essential spectrum of $T(a)$.

Part II

We denote by $\tilde{\mathcal{H}}$ the image of $\mathcal{H}$ under P, and we consider $T(a) = PM(a)P$ as an operator on $\tilde{\mathcal{H}}$, which we denote by $\tilde{T}(a)$ in this case.

(a) Let $a \in C^0(\mathbb{S}^1, \mathbb{C})$ be such that a does not vanish on $\mathbb{S}^1$. Show that $\tilde{T}(a)$ is a Fredholm operator. For this, we suggest that you find an inverse for $\tilde{T}(a)$ modulo a compact operator.

(b) Let $a \in C^0(\mathbb{S}^1, \mathbb{C})$ be such that $a(\theta_0) = 0$. Show that $\tilde{T}(a)$ is not a Fredholm operator.

(c) Consider $a(\theta) = \exp ik\theta$ for a fixed $k \in \mathbb{Z}$. Show that

$$\mathrm{Ind}\,(\tilde{T}(a)) = -k.$$

(d) Show that for $a(\theta) = \exp ik\theta$, we have

$$\mathrm{Ind}\,(\tilde{T}(a)) = -\frac{1}{2\pi} \int_0^{2\pi} \frac{a'(\theta)}{a(\theta)}\, d\theta.$$

(e) Show that if the right-hand side of the previous formula equals $-k$, then there exists a (2π)-periodic function b in C^1 such that

$$a(\theta) = \exp b(\theta) \exp ik\theta.$$

(f) By a deformation argument, deduce that such a formula also holds as long as a is in $C^1(\mathbb{S}^1)$ and does not vanish on $\mathbb{S}^1$.

Problem 16.2 Consider the following problem on $]0, 1[$:

$$(*) \quad -u''(x) = f(x), \quad u(0) = u'(1) = 0.$$

(a) Show that if f is a real-valued continuous function on $[0, 1]$, then there exists a unique solution u in $C^2([0, 1])$ of this problem.

(b) Let T be the operator defined for $f \in \mathcal{H} = L^2(]0, 1[\,;\mathbb{R})$ by

$$(Tf)(x) = \int_0^x f(t)\, dt,$$

and let $\mathbf{K} = T\,T^*$. Show that if $f \in C^0$, then $u = \mathbf{K}f$ is a solution in $C^2([0, 1])$ of $(*)$.

Show that if $f \in L^2$, then $u = \mathbf{K}f$ is a solution in $H^2(]0, 1[)$ of $(*)$.

(c) Show that **K** is positive and compact. Deduce that 0 belongs to the essential spectrum of **K**.

(d) Show that **K** is injective.

(e) Show that if u is an eigenvector of **K** with an eigenvalue $\lambda \neq 0$, then u is a solution of $(*)$ with $f = u/\lambda$. Deduce all the eigenvalues of **K**.

(f) Let q be a strictly positive C^∞ function on $[0, 1]$. Show that there exists a sequence α_n tending to $+\infty$ such that the problem

$$-u'' - \alpha q u = 0, \ u(0) = u'(1) = 0$$

admits, for $\alpha = \alpha_n$, a nontrivial solution in $H^2(]0, 1[)$.

We suggest that you consider the spectrum of $\widehat{K} := q\mathbf{K}$ and show that it is compact and self-adjoint on $L^2(]0, 1[, \rho\, dx)$ for a suitable function $\rho > 0$. (Here $L^2(]0, 1[, \rho\, dx)$ denotes the space of the L^2 functions relative to the measure $\rho(x)\, dx$.)

Problem 16.3 Let Ω be a bounded, regular open set in $\mathbb{C}$. We denote by $\mathcal{H}^2(\Omega)$ the subspace in $L^2(\Omega)$ of the holomorphic L^2 functions in Ω.

(a) Show that $\mathcal{H}^2(\Omega)$ is a closed subspace in $L^2(\Omega)$.

Let Π be the projector of $L^2(\Omega)$ onto $\mathcal{H}^2(\Omega)$ and, for a given function $a \in C^0(\overline{\Omega})$, let T_a be the operator of multiplication by a on $L^2(\Omega)$. We then define P_a on $\mathcal{H}^2(\Omega)$ by $P_a = \Pi\, T_a$.

(b) Show that T_a and P_a are in $\mathcal{L}(L^2(\Omega))$ and $\mathcal{L}(\mathcal{H}^2(\Omega))$, respectively.

(c) Using suitable Cauchy formulas, show that if $D(a_1, r_1)$ and $D(a_2, r_2)$ denote two disks $\overline{D(a_1, r_1)} \subset D(a_2, r_2)$, then there exists, for all $k \in \mathbb{N}$, a constant $C_k > 0$ such that for all f holomorphic in a neighborhood of $\overline{D(a_2, r_2)}$,

$$\sup_{z \in D(a_1, r_1)} |f^{(k)}(z)| \leq C_k \|f\|_{L^2(D(a_2, r_2))}.$$

(d) Show that if $\mathcal{B}$ is bounded in $\mathcal{H}^2(\Omega)$, then, for all compact sets $K \subset \Omega$, the space $\mathcal{B}_K$ obtained by considering the restriction to K of elements in $\mathcal{B}$ is a bounded subset of $C^1(K)$.

(e) Assume now that $a \in C_0^0(\Omega)$, i.e., it is continuous with compact support in Ω. Show that T_a is a compact operator from $\mathcal{H}^2(\Omega)$ into $C^0(\overline{\Omega})$. Deduce that P_a is a compact operator from $\mathcal{H}^2(\Omega)$ into itself.

(f) Show that for any $a \in C^0(\overline{\Omega})$ vanishing at the boundary of Ω, there exists a sequence a_n in $C_0^0(\Omega)$ such that a_n converges to a in $C^0(\overline{\Omega})$.

(g) Show that if $a \in C^0(\overline{\Omega})$ vanishes on $\partial\Omega$, then $P_a \in \mathcal{K}(\mathcal{H}^2(\Omega))$.

(h) Assume now that a is constant and nonzero on $\partial\Omega$. Show that P_a is a Fredholm operator from $\mathcal{H}^2(\Omega)$ into itself.

Problem 16.4 Consider, in the disk $\Omega := D(0, R)$ of $\mathbb{R}^2$, the Dirichlet realization of the Schrödinger operator

$$S(h) := -\Delta + \frac{1}{h^2} V(x), \qquad (16.0.1)$$

where V is a C^∞ potential on $\bar{\Omega}$ satisfying

$$V(x) \geq 0. \qquad (16.0.2)$$

Here $h > 0$ is a parameter.

(a) Show that this operator has a compact resolvent.

(b) Let $\lambda_1(h)$ be the lowest eigenvalue of $S(h)$. We would like to analyze the behavior of $\lambda_1(h)$ as $h \to 0$. Show that $h \to \lambda_1(h)$ is monotonically increasing.

(c) Assume that $V > 0$ on $\bar{\Omega}$. Show that there exists $\epsilon > 0$ such that

$$h^2 \lambda_1(h) \geq \epsilon. \qquad (16.0.3)$$

(d) Assume now that $V = 0$ in an open set ω in Ω. Show that there exists a constant $C > 0$ such that, for any $h > 0$,

$$\lambda_1(h) \leq C. \qquad (16.0.4)$$

To do this, we suggest that you study the Dirichlet realization of $-\Delta$ in ω.

(e) Assume that

$$V > 0 \text{ almost everywhere in } \Omega. \qquad (16.0.5)$$

Show that, under this assumption,

$$\lim_{h \to 0} \lambda_1(h) = +\infty. \qquad (16.0.6)$$

We suggest that you proceed by contradiction, by assuming that there exists C such that

$$\lambda_1(h) \leq C, \ \forall h \text{ such that } 1 \geq h > 0, \qquad (16.0.7)$$

and establishing the following properties.

– For $h > 0$, denote by $x \mapsto u_1(h)(x)$ an L^2-normalized eigenfunction associated with $\lambda_1(h)$. Show that the family $u_1(h)$ $(0 < h \leq 1)$ is bounded in $H^1(\Omega)$.

- Show the existence of a sequence h_n ($n \in \mathbb{N}$) tending to 0 as $n \to +\infty$ and a $u_\infty \in L^2(\Omega)$ such that

$$\lim_{n \to +\infty} u_1(h_n) = u_\infty$$

in $L^2(\Omega)$.
- Deduce that

$$\int_\Omega V(x)u_\infty(x)^2 \, dx = 0.$$

- Deduce that $u_\infty = 0$, and make the contradiction explicit.

(f) Assume that $V(0) = 0$. Show that there exists a constant C such that

$$\lambda_1(h) \leq \frac{C}{h}.$$

(g) Assume that $V(x) = \mathcal{O}(|x|^4)$ near 0. Show that in this case

$$\lambda_1(h) \leq \frac{C}{h^{2/3}}.$$

(h) Assume that $V(x) \sim |x|^2$ near 0. Discuss whether one can hope for a lower bound in the form

$$\lambda_1(h) \geq \frac{1}{Ch}.$$

Justify the answer by illustrating the arguments with examples and counterexamples.

Problem 16.5 Consider the self-adjoint operator $H_\epsilon = -d^2/dx^2 + x^2 + \epsilon |x|$ on $\mathbb{R}$, for $\epsilon \in I := [-\frac{1}{4}, +\infty[$.

(a) Determine the form domain of H_ϵ and show that it is independent of ϵ.
(b) What is the nature of the spectrum of the associated self-adjoint operator?
(c) Let $\lambda_1(\epsilon)$ be the smallest eigenvalue. Give rough estimates that permit one to estimate $\lambda_1(\epsilon)$ from above or below, independently of ϵ on every compact interval of I.
(d) Show that for any compact subinterval J of I, there exists a constant C_J such that for all $\epsilon \in J$, any L^2-normalized eigenfunction u_ϵ of H_ϵ associated with $\lambda_1(\epsilon)$ satisfies $\|u_\epsilon\|_{B^1(\mathbb{R})} \leq C_J$. For this, we suggest that you play with $\langle H_\epsilon u_\epsilon, u_\epsilon \rangle_{L^2(\mathbb{R})}$.

(e) Show that the lowest eigenvalue is a monotonically increasing sequence of $\epsilon \in I$.

(f) Show that the lowest eigenvalue is a locally Lipschitzian function of $\epsilon \in I$. (We suggest that you use the max–min principle again.)

(g) Show that $\lambda(\epsilon) \to +\infty$ as $\epsilon \to +\infty$, and estimate the asymptotic behavior.

(h) Discuss the same questions for the case $H_\epsilon = -d^2/dx^2 + x^2 + \epsilon x^4$ (with $\epsilon \geq 0$).

Problem 16.6 The aim of this problem is to analyze the spectrum $\Sigma^D(P)$ of the Dirichlet realization of the operator $P := (D_{x_1} - \frac{1}{2}x_2)^2 + (D_{x_2} + \frac{1}{2}x_1)^2$ (the magnetic Laplacian) in $\mathbb{R}^+ \times \mathbb{R}$.

(a) Show that one can compare, a priori, the infimum of the spectrum of P in $\mathbb{R}^2$ and the infimum of $\Sigma^D(P)$.

(b) Compare $\Sigma^D(P)$ with the spectrum $\Sigma^D(Q)$ of the Dirichlet realization of $Q := D_{y_1}^2 + (y_1 - y_2)^2$ in $\mathbb{R}^+ \times \mathbb{R}$.

(c) Consider first the following family of Dirichlet problems associated with the family of differential operators $\alpha \mapsto H(\alpha)$ defined on $]0, +\infty[$ by

$$H(\alpha) = D_t^2 + (t - \alpha)^2.$$

Compare this family with the Dirichlet realization of the harmonic oscillator in $]-\alpha, +\infty[$.

(d) Show that the lowest eigenvalue $\lambda(\alpha)$ of $H(\alpha)$ is a monotonic function of $\alpha \in \mathbb{R}$.

(e) Show that $\alpha \mapsto \lambda(\alpha)$ is a continuous function on $\mathbb{R}$.

(f) Analyze the limit of $\lambda(\alpha)$ as $\alpha \to -\infty$.

(g) Analyze the limit of $\lambda(\alpha)$ as $\alpha \to +\infty$.

(h) Compute $\lambda(0)$. For this, we suggest that you compare the spectrum of $H(0)$ with the spectrum of the harmonic oscillator restricted to odd functions.

(i) Let $t \mapsto u(t; \alpha)$ be the positive L^2-normalized eigenfunction associated with $\lambda(\alpha)$. Assume that this is the restriction to $\mathbb{R}^+$ of a function in $S(\mathbb{R})$. Let T_α be the distribution in $\mathcal{D}'(\mathbb{R}^+ \times \mathbb{R})$ defined by

$$\phi \mapsto T_\alpha(\phi) = \int_0^{+\infty} \phi(y_1, \alpha)u_\alpha(y_1)\,dy_1,$$

for $\alpha \in \mathbb{R}$. Compute QT_α.

(j) By constructing, starting from T_α, a suitable sequence of L^2 functions tending to T_α, show that $\lambda(\alpha) \in \Sigma^D(Q)$.

(k) Determine $\Sigma^D(P)$.

Problem 16.7 Let H_a be the Dirichlet realization of $-d^2/dx^2 + x^2$ in $]-a, +a[$.

(a) Briefly recall the results concerning the case $a = +\infty$.

(b) Show that the lowest eigenvalue $\lambda_1(a)$ of H_a is decreasing for $a \in]0, +\infty[$ and larger than 1.

(c) Show that $\lambda_1(a)$ tends exponentially fast to 1 as $a \to +\infty$. (We suggest that you use a suitable construction of approximate eigenvectors.)

(d) What is the behavior of $\lambda_1(a)$ as $a \to 0$? (We suggest that you use the change of variable $x = ay$ and analyze the limit $\lim_{a \to 0} a^2 \lambda_1(a)$.)

(e) Let $\mu_1(a)$ be the smallest eigenvalue of the Neumann realization in $]-a, +a[$. Show that $\mu_1(a) \le \lambda_1(a)$.

(f) Show that if u_a is a normalized eigenfunction associated with $\mu_1(a)$, then there exists a constant C such that, for all $a \ge 1$, we have

$$\|xu_a\|_{L^2(]-a,+a[)} \le C.$$

(g) Show that for u in $C^2([-a, +a])$ and χ in $C_0^2(]-a, +a[)$, we have

$$-\int_{-a}^{+a} \chi(t)^2 u''(t)u(t)\,dt = \int_{-a}^{+a} |(\chi u)'(t)|^2\,dt - \int_{-a}^{+a} \chi'(t)^2 u(t)^2\,dt.$$

(h) Using this identity with $u = u_a$, a suitable χ (which should be equal to 1 on $[-a+1, a-1]$), the estimate obtained in (f), and the minimax principle, show that there exists C such that for $a \ge 1$ we have

$$\lambda_1(a) \le \mu_1(a) + Ca^{-2}.$$

Deduce the limit of $\mu_1(a)$ as $a \to +\infty$.

Problem 16.8 (Semiclassical analysis and the Airy operator.) We wish to understand the problem on $\mathbb{R}^+$ given by the Dirichlet realization $P^D(h)$ of

$$P(h) := -h^2 \frac{d^2}{dx^2} + v(x),$$

with $v'(x) \ge c > 0$ on $\overline{\mathbb{R}^+}$.

(a) Show that this operator has a compact resolvent.

(b) First, analyze the case $v(x) = x$, $h = 1$. (In this case, the operator is called the Airy operator and denoted by $A(x, D_x)$.) Show that, for the Dirichlet realization A^D of A in $\mathbb{R}^+$, there exists a sequence $(\mu_j)_{j \in \mathbb{N}^*}$ of eigenvalues tending to ∞. Show that the lowest one, μ_1, is strictly positive. What is the form domain $Q(A^D)$ of the Airy operator?

(c) Show that the corresponding eigenfunctions u_j are in $C^\infty(\overline{\mathbb{R}^+})$.

(d) Show that the eigenvalues are of multiplicity 1.

(e) Assume that

$$D(A^D) = \{u \in H_0^1(\mathbb{R}^+) \cap H^2(\mathbb{R}^+); xu \in L^2(\mathbb{R}^+)\}$$
$$= \{u \in H_0^1(\mathbb{R}^+), x^{1/2}u \in L^2(\mathbb{R}^+), A(x, D_x)u \in L^2(\mathbb{R}^+)\}.$$

Show that the eigenvectors are in $\mathcal{S}(\overline{\mathbb{R}^+})$.

Another approach could be to analyze the Fourier transform of χu_j, where χ is equal to 1 for x large and equal to 0 in a neighborhood of 0.

(f) Describe the spectrum of $A^D(x, hD_x)$ for any $h > 0$.

(g) We now return to the general case. Transpose what was done for the one-well problem via the harmonic approximation to $P^D(h)$, where the harmonic oscillator is replaced by the Airy operator. We suggest that you use, if necessary, the fact that $(A^D(x, D_x) - \mu_1)$ is a bijection from $\mathcal{S}_0(\overline{\mathbb{R}^+}) \cap \{\mathbb{R}u_1\}^\perp$ onto $\mathcal{S}(\overline{\mathbb{R}^+}) \cap \{\mathbb{R}u_1\}^\perp$, where

$$\mathcal{S}_0(\overline{\mathbb{R}^+}) = \{u \in \mathcal{S}(\overline{\mathbb{R}^+}) \text{ s.t. } u(0) = 0\}.$$

Problem 16.9 (Schrödinger operator in $\mathbb{R}_+^2$ with Dirichlet conditions.) The aim of this problem is to analyze the spectrum $\Sigma^D(P)$ of the Dirichlet realization of the operator $P := (D_{x_1} - \frac{1}{2}x_2)^2 + (D_{x_2} + \frac{1}{2}x_1)^2$ in $\mathbb{R}^+ \times \mathbb{R}$.

(a) Show that one can compare, a priori, the infimum of the spectrum of P in $\mathbb{R}^2$ and the infimum of $\Sigma^D(P)$.

(b) Compare $\Sigma^D(P)$ with the spectrum $\Sigma^D(Q)$ of the Dirichlet realization of $Q := D_{y_1}^2 + (y_1 - y_2)^2$ in $\mathbb{R}^+ \times \mathbb{R}$.

First, consider the family of Dirichlet problems associated with the family of differential operators $\alpha \mapsto H(\alpha)$ defined on $]0, +\infty[$ by

$$H(\alpha) = D_t^2 + (t - \alpha)^2.$$

Compare this family with the Dirichlet realization of the harmonic oscillator in $]-\alpha, +\infty[$.

(c) Show that the lowest eigenvalue $\lambda(\alpha)$ of $H(\alpha)$ is a monotonic function of $\alpha \in \mathbb{R}$.

(d) Show that $\alpha \mapsto \lambda(\alpha)$ is a continuous function on $\mathbb{R}$.

(e) Analyze the limit of $\lambda(\alpha)$ as $\alpha \to -\infty$.

(f) Analyze the limit of $\lambda(\alpha)$ as $\alpha \to +\infty$.

(g) Compute $\lambda(0)$. For this, compare the spectrum of $H(0)$ with the spectrum of the harmonic oscillator restricted to odd functions.

(h) Let $t \mapsto u(t; \alpha)$ be the positive L^2-normalized eigenfunction associated with $\lambda(\alpha)$. Assume that this is the restriction to $\mathbb{R}^+$ of a function in $S(\mathbb{R})$. Let T_α be the distribution in $\mathcal{D}'(\mathbb{R}^+ \times \mathbb{R})$ defined by

$$\phi \mapsto T_\alpha(\phi) = \int_0^{+\infty} \phi(y_1, \alpha) u_\alpha(y_1) \, dy_1,$$

for $\alpha \in \mathbb{R}$. Compute QT_α.

(i) By constructing, starting from T_α, a suitable sequence of L^2 functions tending to T_α, show that $\lambda(\alpha) \in \Sigma^D(Q)$.

(j) Determine $\Sigma^D(P)$.

Problem 16.10 This problem comes from superconductivity theory [Alm]. We consider, for $c \in \mathbb{R}$, the differential operator defined on $C_0^\infty(\mathbb{R}^2)$ by

$$A_{0,c} = D_x^2 + \left(D_y - \frac{1}{2}x^2\right)^2 + icy,$$

considered as an unbounded operator on $L^2(\mathbb{R}^2)$. We recall that $D_x = -i\partial_x$ and $D_y = -i\partial_y$. The problem is about analyzing the spectral properties of $A_{0,c}$ as a function of $c \in \mathbb{R}$.

Part I

For $\eta \in \mathbb{R}$, consider the unbounded operator $\mathfrak{h}_0(\eta)$ on $L^2(\mathbb{R})$, with domain $C_0^\infty(\mathbb{R})$, associated with the differential operator

$$\mathfrak{h}_0(\eta) := -\frac{d^2}{dx^2} + \left(\eta - \frac{x^2}{2}\right)^2.$$

(a) Show that, for any η, one can construct a self-adjoint extension of $\mathfrak{h}_0(\eta)$, and describe its domain.

(b) Show that the operator has a compact resolvent, and deduce that the spectrum consists of a sequence of eigenvalues $\lambda_j(\eta)$ ($j \in \mathbb{N}^*$) tending to $+\infty$.

(c) Show that $\lim_{\eta \to -\infty} \lambda_1(\eta) = +\infty$.

(d) Show that $\lambda_1(\eta) > 0$.

(e) Show that $\eta \to \lambda_1(\eta)$ is continuous.

(f) Show that $\eta \mapsto \lambda_j(\eta)$ is monotonically decreasing for $\eta < 0$.

(g) Assume that $\lim_{\eta \to +\infty} \lambda_1(\eta) = +\infty$. Show that $\eta \mapsto \lambda_1(\eta)$ attains its infimum λ^* at at least one point.

(h) Assume that $\eta \mapsto \lambda_1(\eta)$ is of multiplicity 1 and class C^1 and that one can, for any η, associate with $\lambda_1(\eta)$ an eigenfunction $u_1(\cdot, \eta)$ in $S(\mathbb{R})$ with a C^1 dependence of η such that $\|u_1\| = 1$ and $u_1 > 0$. Show that

$$\lambda_1'(\eta) = 2 \int_{\mathbb{R}} \left(\eta - \frac{x^2}{2} \right) u_1(x, \eta)^2 \, dx.$$

Deduce that the critical points η_c of λ_1 satisfy $\eta_c > 0$.

(i) Show that $u_1(\cdot, \eta)$ is even.

(j) Show that if η_c is a critical point of λ_1, then

$$I(\eta_c) := \int_0^{+\infty} x \left(\frac{x^2}{2} - \eta_c \right) u_1(x, \eta_c)^2 \, dx \geq 0.$$

(k) By computing $I(\eta)$ differently, deduce that

$$\eta_c^2 \leq \lambda_1(\eta_c).$$

(l) Using Gaussian quasimodes for $\mathfrak{h}_1(0)$, determine an interval (as good as possible) in which λ_1 has its minimum.

Part II

Here we suppose that $c = 0$, and write A_0 for $A_{0,0}$.

(a) Show that the operator A_0 is symmetric.

(b) Show that one can construct its Friedrichs extension A_0^{Fried}, associated with a sesquilinear form, to be defined precisely. Describe the form domain and the domain of the operator.

(c) Show that the operator is essentially self-adjoint.

(d) Show that

$$\sigma(A_0^{\text{Fried}}) \subset [\lambda^*, +\infty[.$$

(We suggest that you use a partial Fourier transform with respect to y.)

(e) By constructing suitable families of approximate eigenfunctions, show that

$$\sigma(A_0^{\text{Fried}}) = [\lambda^*, +\infty[.$$

(f) Show that A_0^{Fried} does not have a compact resolvent.

Part III

We now suppose that $c \neq 0$.

(a) Show that

$$\operatorname{Re}\langle A_{0,c}u,\, u\rangle = \langle A_0 u,\, u\rangle$$

and

$$\operatorname{Im}\langle A_{0,c}u,\, u\rangle = c\,\langle yu,\, u\rangle,$$

for all $u \in D(A_{0,c})$.

(b) Show that

$$\operatorname{Re}\langle A_{0,c}u,\, u\rangle \geq \lambda^*\,||u||^2,$$

for all $u \in D(A_{0,c})$.

(c) Show that $A_{0,c}$ is closable. We denote its closure by $\overline{A_{0,c}}$. Recall how this operator is defined, and describe its domain.

(d) Show that for $\lambda > -\lambda^*$, $\overline{A_{0,c}} + \lambda$ is injective and has a closed range.

(e) Show that for $\lambda > -\lambda^*$, $\overline{A_{0,c}} + \lambda$ has a dense range. (We suggest that you adapt the proof of item (c) in Part II.)

(f) Show that

$$\sigma(\overline{A_{o,c}}) \subset \{\lambda \in \mathbb{C},\ \operatorname{Re}\lambda \geq \lambda^*\}.$$

The aim of questions (g)–(k) below is to show that $\overline{A_{o,c}}$ has a compact resolvent, for $c \neq 0$.

(g) Show that for any compact set $K \subset \mathbb{R}^2$, there exists a constant C_K such that, for any $u \in C_0^\infty(\mathbb{R}^2)$ with support in K, we have

$$||u||^2_{H^1(\mathbb{R}^2)} \leq C_K \left(\operatorname{Re}\langle A_{0,c}u,\, u\rangle + ||u||^2_{L^2} \right).$$

(h) Show that for all $u \in C_0^\infty(\mathbb{R}^2)$ with support in $\{y > 0\}$ or in $\{y < 0\}$, we have

$$\int |y|\, |u(x, y)|^2\, dx\, dy \leq \frac{1}{|c|}\, |\operatorname{Im}\langle A_{0,c}u,\, u\rangle|.$$

By using a partition of unity, deduce that

$$||\, |y|^{1/2}u||^2 \leq C \left(||A_0 u||^2_{L^2} + ||u||^2_{L^2} \right).$$

(i) Show that for all $u \in C_0^\infty(\mathbb{R}^2)$ with support in $\{x > 0\}$ or in $\{x < 0\}$, we have

$$\int |x|\,|u(x, y)|^2\, dx\, dy \le \operatorname{Re}\langle A_{0,c}u, u\rangle.$$

By using a partition of unity, deduce that there exists a constant C such that, for all $u \in C_0^\infty(\mathbb{R}^2)$, we have

$$\| |x|^{1/2}u\|^2 \le C\left(\|A_0 u\|_{L^2}^2 + \|u\|_{L^2}^2\right).$$

(j) Deduce that $\overline{A_{0,c}}$ has a compact resolvent.
(k) Show that if (u, λ) is a spectral pair for $\overline{A_{0,c}}$, then, for all $a \in \mathbb{R}$, the pair $(u_a, \lambda - ica)$, where u_a is defined by $u_a(x, y) = u(x, y + a)$, is spectral.
(l) Deduce that the spectrum of $\overline{A_{0,c}}$ is empty.
(m) Show that $\overline{A_{0,c}} = A_{0,-c}^*$.

Problem 16.11 (After [Ris] and [HelN].) Consider the Kramers–Fokker–Planck operator on $L^2(\mathbb{R}^2)$ introduced in Chapter 15,

$$L = -\partial_v^2 + \frac{1}{4}v^2 - \frac{1}{2} - \tilde{\omega}_0(v\partial_x - x\partial_v),$$

with $\tilde{\omega}_0 \ne 0$.

(a) With

$$b = \partial_v + \frac{1}{2}v, \quad a = \partial_x + \frac{1}{2}x,$$

show that this operator can also be written as

$$L = b^*b + \tilde{\omega}_0(b^*a - a^*b).$$

(b) We would like to rewrite L as a "complex" harmonic oscillator. Show that we can write

$$L = \lambda_1 c_{1,+}c_{1,-} + \lambda_2 c_{2,+}c_{2,-},$$

where λ_1 and λ_2 are complex numbers, and $c_{1,-}$, $c_{1,+}$, $c_{2,-}$, and $c_{2,+}$ satisfy the standard commutation relations

$$[c_{1,-}, c_{1,+}] = [c_{2,-}, c_{2,+}] = 1,$$
$$[c_{1,-}, c_{2,+}] = [c_{1,+}, c_{2,-}] = [c_{1,-}, c_{2,-}] = [c_{1,+}, c_{2,+}] = 0,$$

and another equation,

$$[L, \ c_{i,\pm}] = \mp c_{i,\pm}.$$

To do this, look for $c_{j,\pm}$ in the form

$$
\begin{aligned}
c_{1,+} &= \delta^{-1/2} \left(\sqrt{\lambda_1} b^* - \sqrt{\lambda_2} a^* \right), \\
c_{1,-} &= \delta^{-1/2} \left(\sqrt{\lambda_1} b + \sqrt{\lambda_2} a \right), \\
c_{2,+} &= \delta^{-1/2} \left(-\sqrt{\lambda_2} b^* + \sqrt{\lambda_1} a^* \right), \\
c_{2,-} &= \delta^{-1/2} \left(\sqrt{\lambda_2} b + \sqrt{\lambda_1} a \right),
\end{aligned}
$$

where

$$\delta = \sqrt{1 - 4\tilde{\omega}_0^2}$$

is a chosen square root of $1 - 4\tilde{\omega}_0^2$ (which is assumed to be different from 0) and

$$\lambda_1 = \frac{1 + \delta}{2}, \ \lambda_2 = \frac{1 - \delta}{2}.$$

(c) Is $c_{j,-}$ the formal adjoint of $c_{j,+}$?

(d) Show that the construction of the eigenvectors by use of the "creation operator" works and that one can obtain a complete system of eigenvectors by introducing

$$\psi_{n_1,n_2} = (n_1! \, n_2!)^{-1/2} (c_{1,+})^{n_1} (c_{2,+})^{n_2} \psi_{0,0},$$

where

$$\psi_{0,0} = \left(\frac{1}{2\pi} \right)^{1/2} \exp -\frac{1}{4}(x^2 + v^2).$$

The corresponding eigenvalue is

$$\lambda_{n_1,n_2} = \lambda_1 n_1 + \lambda_2 n_2.$$

(e) Discuss the properties of the eigenvalues as a function of δ.

Bibliography

[ABHN]   Arendt, W., Batty, C., Hieber, M., and Neubrander, F. 2001. *Vector-Valued Laplace Transforms and Cauchy Problems*. Birkhäuser.

[AbSt]   Abramowitz, M. and Stegun, I. A. 1964. *Handbook of Mathematical Functions*, Applied Mathematics Series, Vol. 55. National Bureau of Standards.

[Ag]   Agmon, S. 1982. *Lecture on Exponential Decay of Solutions of Second Order Elliptic Equations*, Mathematical Notes, Vol. 29. Princeton University Press.

[AHP1]   Almog, Y., Helffer, B., and Pan, X. 2010. Superconductivity near the normal state under the action of electric currents and induced magnetic field in $\mathbb{R}^2$. *Commun. Math. Phys.*, **300**(1), 147–184.

[AHP2]   Almog, Y., Helffer, B., and Pan, X. 2011. Superconductivity near the normal state in a half-plane under the action of a perpendicular electric current and an induced magnetic field. To appear in *Trans. Am. Math. Soc.*

[AHS]   Avron, J., Herbst, I., and Simon, B. 1978. Schrödinger operators with magnetic fields I. *Duke Math. J.*, **45**, 847–883.

[AkGl]   Akhiezer, N. I. and Glazman, I. M. 1981. *Theory of Linear Operators in Hilbert Space*. Pitman.

[Alm]   Almog, Y. 2008. The stability of the normal state of superconductors in the presence of electric currents. *SIAM J. Math. Anal.*, **40**(2), 824–850.

[AsDa]   Aslayan, A. and Davies, E. B. 2000. Spectral instability for some Schrödinger operators. *Numer. Math.*, **85**, 525–552.

[BeSt]   Bernoff, A. and Sternberg, P. 1998. Onset of superconductivity in decreasing fields for general domains. *J. Math. Phys.*, **39**, 1272–1284.

[BGM]   Berger, M., Gauduchon, P., and Mazet, E. 1971. *Spectre d'une Variété Riemannienne*, Lecture Notes in Mathematics, Vol. 194. Springer.

[BLP]   Benguria, R., Levitin, M., and Parnovski, L. 2009. Fourier transform, null variety, and Laplacian's eigenvalues. *J. Funct. Anal.*, **257**(7), 2088–2123.

[BM]   Bordeaux-Montrieux, W. 2010. Estimation de résolvante et construction de quasi-modes près du bord du pseudospectre. Preprint.

[BoKr]   Borisov, D. and Krejcirik, D. 2012. The effective Hamiltonian for thin layers with non-hermitian Robin-type boundary. *Asymptot. Anal.*, **76**, 49–59.

[Bou]   Boulton, L. S. 2002. Non-self-adjoint harmonic oscillator, compact semigroups and pseudospectra. *J. Oper. Theory*, **47**(2), 413–429.

[Br]   Brézis, H. 2005. *Analyse Fonctionnelle*. Editions Masson.

[BS]   Blanchard, P. and Stubbe, J. 1996. Bound states for Schrödinger Hamiltonians. Phase space methods and applications. *Rev. Math. Phys.*, **35**, 504–547.

[CCLaRa] Cherfils-Clerouin, C., Lafitte, O., and Raviart, P.-A. 2001. Asymptotics results for the linear stage of the Rayleigh–Taylor instability. In Neustupa, J. and Penel, P. (eds.),

	Mathematical Fluid Mechanics: Recent Results and Open Questions, Advances in Mathematical Fluid Mechanics, pp. 47–71. Birkhäuser.
[CDV]	Colin de Verdière, Y. 1998. *Spectres de graphes*, Cours Spécialisé, 4. Société Mathématique de France.
[CFKS]	Cycon, H. L., Froese, R., Kirsch, W., and Simon, B. 1987. *Schrödinger Operators: With Applications to Quantum Mechanics and Global Geometry*, Texts and Monographs in Physics. Springer.
[CH]	Courant, R. and Hilbert, D. 1953. *Methods of Mathematical Physics*. Wiley-Interscience.
[ChLa]	Cherfils, C. and Lafitte, O. 2000. Analytic solutions of the Rayleigh equation for linear density profiles. *Phys. Rev. E*, **62**(2), 2967–2970.
[CT]	Combes, J.-M. and Thomas, L. 1973. Asymptotic behaviour of eigenfunctions for multiparticle Schrödinger operators. *Commun. Math. Phys.*, **34**, 251–270.
[DaHe]	Dauge, M. and Helffer, B. 1993. Eigenvalues variation I, Neumann problem for Sturm–Liouville operators. *J. Differ. Equ.*, **104**(2), 243–262.
[DaLi]	Dautray, R. and Lions, J.-L. 1988–1995. *Analyse Mathématique et Calcul Numérique pour les Sciences et les Techniques*. Masson.
[Dav1]	Davies, E. B. 1996. *Spectral Theory and Differential Operators*, Cambridge Studies in Advanced Mathematics, Vol. 42. Cambridge University Press.
[Dav2]	Davies, E. B. 1999. Semi-classical states for non-self-adjoint Schrödinger operators. *Commun. Math. Phys.*, **200**, 35–41.
[Dav3]	Davies, E. B. 1999. Pseudospectra, the harmonic oscillator and complex resonances. *Proc. R. Soc. Lond. Ser. A*, **455**, 585–599.
[Dav4]	Davies, E. B. 2000. Wild spectral behaviour of anharmonic oscillators. *Bull. Lond. Math. Soc.*, **32**, 432–438.
[Dav5]	Davies, E. B. 2002. Non-self-adjoint differential operators. *Bull. Lond. Math. Soc.*, **34**, 513–532.
[Dav6]	Davies, E. B. 2005. Semigroup growth bounds. *J. Oper. Theory*, **53**(2), 225–249.
[Dav7]	Davies, E. B. 2007. *Linear Operators and Their Spectra*, Cambridge Studies in Advanced Mathematics, Vol. 106. Cambridge University Press.
[Di]	Dieudonné, J. 1980. *Calcul Infinitésimal*. Hermann.
[DiSj]	Dimassi, M. and Sjöstrand, J. 1999. *Spectral Asymptotics in the Semi-Classical Limit*, London Mathematical Society Lecture Note Series, Vol. 268. Cambridge University Press.
[DSZ]	Dencker, N., Sjöstrand, J., and Zworski, M. 2004. Pseudospectra of semi-classical (pseudo)differential operators. *Commun. Pure Appl. Math.*, **57**(3), 384–415.
[DW]	Deng, W. 2012. *Etude du pseudo-spectre d'opérateurs non auto-adjoints liés à la mécanique des fluides*. Thèse de doctorat, Université Pierre et Marie Curie.
[EcPRB]	Eckmann, J.-P., Pillet, C. A., and Rey-Bellet, L. 1999. Non-equilibrium statistical mechanics of anharmonic chains coupled to two heat baths at different temperatures. *Commun. Math. Phys.*, **208**(2), 275–281.
[EnNa1]	Engel, K. J. and Nagel, R. 2000. *One-Parameter Semigroups for Linear Evolution Equations*, Graduate Texts in Mathematics, Vol. 194. Springer.
[EnNa2]	Engel, K. J. and Nagel, R. 2005. *A Short Course on Operator Semi-Groups*, Unitext. Springer.
[FoHe]	Fournais, S. and Helffer, B. 2010. *Spectral Methods in Surface Superconductivity*, Progress in Nonlinear Differential Equations and Their Applications, Vol. 77. Birkhäuser.
[Ge]	Gearhart, L. 1978. Spectral theory for contraction semigroups on Hilbert spaces. *Trans. Am. Math. Soc.*, **236**, 385–394.
[GGN]	Gallagher, I., Gallay, T., and Nier, F. 2009. Spectral asymptotics for large skew-symmetric perturbations of the harmonic oscillator. *Int. Math. Res. Not.*, **12**(12), 2147–2199.

[GiT]   Gilbarg, D. and Trudinger, N. S. 1998. *Elliptic Partial Differential Equations of Second Order.* Springer.

[GlJa]   Glimm, J. and Jaffe, A. 1987. *Quantum Physics: A Functional Integral Point of View,* 2nd edition. Springer.

[GrSj]   Grigis, A. and Sjöstrand, J. 1994. *Microlocal Analysis for Differential Operators: An Introduction,* London Mathematical Society Lecture Note Series, Vol. 196. Cambridge University Press.

[Hag]   Hager, M. 2006. Instabilité spectrale semi-classique pour des opérateurs non-autoadjoints I: un modèle. *Ann. Fac. Sci. Toulouse Math.* (6), **15**(2), 243–280.

[Har]   Hardy, G. H. 1920. Note on a theorem of Hilbert. *Math. Z.,* **6**, 314–317.

[He1]   Helffer, B. 1984. *Théorie Spectrale pour des Opérateurs Globalement Elliptiques,* Astérisque, Vol. 112. Société Mathématique de France.

[He2]   Helffer, B. 1988. *Semiclassical Analysis for the Schrödinger Operator and Applications,* Lecture Notes in Mathematics, Vol. 1336. Springer.

[He3]   Helffer, B. 1995. *Semiclassical Analysis for Schrödinger Operators, Laplace Integrals and Transfer Operators in Large Dimension: An Introduction,* Cours de DEA. Paris Onze Edition.

[He4]   Helffer, B. 2002. *Semiclassical Analysis, Witten Laplacians and Statistical Mechanics,* Series on Partial Differential Equations and Applications, Vol. 1. World Scientific.

[He7]   Helffer, B. 2011. On pseudo-spectral problems related to a time dependent model in superconductivity with electric current. *Confluentes Math.,* **3**(2), 237–251.

[HelLaf]   Helffer, B. and Lafitte, O. 2003. Asymptotics methods for the eigenvalues of the Rayleigh equation. *Asymptot. Anal.,* **23**(3–4), 189–236.

[HelN]   Helffer, B. and Nier, F. 2004. *Hypoelliptic Estimates and Spectral Theory for Fokker–Planck Operators and Witten Laplacians,* Lecture Notes in Mathematics, Vol. 1862. Springer.

[HelNo]   Helffer, B. and Nourrigat, J. 1985. *Hypoellipticité Maximale pour des Opérateurs Polynômes de Champs de Vecteurs,* Progress in Mathematics, Vol. 58. Birkhäuser.

[Hen]   Henry, R. 2010. Master's thesis, Université Paris-Sud 11.

[HeSj1]   Helffer, B. and Sjöstrand, J. 1984. Multiple wells in the semiclassical limit I. *Commun. Partial Differ. Equ.,* **9**(4), 337–408.

[HeSj2]   Helffer, B. and Sjöstrand, J. 2010. From resolvent bounds to semigroup bounds. Preprint, arXiv:1001.4171v1.

[Her]   Herbst, I. 1979. Dilation analyticity in constant electric field I. The two body problem. *Commun. Math. Phys.,* **64**, 279–298.

[HerN]   Hérau, F. and Nier, F. 2004. Isotropic hypoellipticity and trend to equilibrium for the Fokker–Planck equation with a high-degree potential. *Arch. Ration. Mech. Anal.,* **171**(2), 151–218.

[HerSjSt] Hérau, F., Sjöstrand, J., and Stolk, C. 2005. Semi-classical analysis for the Kramers–Fokker–Planck equation. *Commun. Partial Differ. Equ.,* **30**(5–6), 689–760.

[HFL]   Huang, F. L. 1985. Characteristic conditions for exponential stability of linear dynamical systems in Hilbert spaces. *Ann. Differ. Equ.,* **1**, 43–56.

[HHS1]   Hérau, F., Hitrik, M., and Sjöstrand, J. 2008. Tunnel effect for Kramers–Fokker–Planck type operators. *Ann. Henri Poincaré,* **9**(2), 209–274.

[HHS2]   Hérau, F., Hitrik, M., and Sjöstrand J. 2008. Kramers–Fokker–Planck type operators: return to equilibrium and applications. *Int. Math. Res. Not.,* Article ID rnn057, 48 pp.

[HiSi]   Hislop, P. D. and Sigal, I. M. 1995. *Introduction to Spectral Theory: With Applications to Schrödinger Operators,* Applied Mathematical Sciences, Vol. 113. Springer.

[Ho1]   Hörmander, L. 1967. Hypoelliptic second order differential equations. *Acta Math.,* **119**, 147–171.

[Ho2]   Hörmander, L. 1985. *The Analysis of Linear Partial Differential Operators.* Springer.

[Hu]   Huet, D. 1976. *Décomposition Spectrale et Opérateurs.* Presses universitaires de France.

[Ka]	Kato, T. 1966. *Perturbation Theory for Linear Operators*. Springer.
[Laf]	Lafitte, O. 2001. Sur la phase linéaire de l'instabilité de Rayleigh-Taylor. Séminaire EDP de l'Ecole Polytechnique. www.math.polytechnique.fr.
[LanTr]	Langer, H. and Tretter, C. 1997. Spectral properties of the Orr–Sommerfeld problem. *Proc. R. Soc. Edinb. Sect. A*, **127**, 1245–1261.
[Lap]	Laptev, A. 1997. Dirichlet and Neumann eigenvalue problems on domains in Euclidean spaces. *J. Funct. Anal.*, **151**(2), 531–545.
[Le-Br]	Lévy-Bruhl, P. 2003. *Introduction à la Théorie Spectrale*. Editions Dunod.
[LiLo]	Lieb, E. and Loss, M. 1996. *Analysis*, Graduate Studies in Mathematics, Vol. 14. American Mathematical Society.
[LiMa]	Lions, J.-L. and Magenes, E. 1968. *Problèmes aux Limites Non-homogènes. Tome 1.* Editions Dunod.
[Lio1]	Lions, J.-L. 1957. *Lecture on Elliptic Partial Differential Equations*. Tata Institute of Fundamental Research, Bombay.
[Lio2]	Lions, J.-L. 1962. Problèmes aux limites dans les EDP. Séminaire de Mathématiques supérieures de l'université de Montréal.
[LiYa]	Li, P. and Yau, S. T. 1983. On the Schrödinger equation and the eigenvalue problem. *Commun. Math. Phys.*, **88**(3), 309–318.
[Mart]	Martinet, J. 2009. *Sur les propriétés spectrales d'opérateurs non-autoadjoints provenant de la mécanique des fluides*. Thèse de doctorat, Université Paris-Sud 11.
[Paz]	Pazy, A. 1983. *Semigroups of Linear Operators and Applications to Partial Differential Operators*, Applied Mathematical Sciences, Vol. 44. Springer.
[Pr]	Prüss, J. 1984. On the spectrum of C_0-semigroups. *Trans. Am. Math. Soc.*, **284**, 847–857.
[PS]	Pravda-Starov, K. 2006. A complete study of the pseudo-spectrum for the rotated harmonic oscillator. *J. Lond. Math. Soc.*, **73**(3), 745–761.
[Ris]	Risken, H. 1989. *The Fokker–Planck Equation: Methods of Solution and Applications*, 2nd edition. Springer.
[Ro]	Robert, D. 1987. *Autour de l'Approximation Semi-classique*, Progress in Mathematics, Vol. 68. Birkhäuser.
[RoSi]	Roch, S. and Silbermann, B. 1996. C^*-algebras techniques in numerical analysis. *J. Oper. Theory*, **35**, 241–280.
[RS-I]	Reed, M. and Simon, B. 1972. *Methods of Modern Mathematical Physics, Vol. I: Functional Analysis*. Academic Press.
[RS-II]	Reed, M. and Simon, B. 1975. *Methods of Modern Mathematical Physics, Vol. II: Fourier Analysis, Self-Adjointness*. Academic Press.
[RS-III]	Reed, M. and Simon, B. 1976. *Methods of Modern Mathematical Physics, Vol. III: Scattering Theory*. Academic Press.
[RS-IV]	Reed, M. and Simon, B. 1978. *Methods of Modern Mathematical Physics, Vol. IV: Analysis of Operators*. Academic Press.
[RSZ]	Rubinstein, J., Sternberg, P., and Zumbrun, K. 2010. The resistive state in a superconducting wire: bifurcation from the normal state. *Arch. Ration. Mech. Anal.*, **195**(1), 117–158.
[Ru1]	Rudin, W. 1974. *Real and Complex Analysis*. McGraw-Hill.
[Ru2]	Rudin, W. 1997. *Analyse Fonctionnelle*. Ediscience International.
[Si1]	Simon, B. 1979. *Functional Integration and Quantum Physics*, Pure and Applied Mathematics, Vol. 86. Academic Press.
[Si2]	Simon, B. 2005. *Trace Ideals and Their Applications*, 2nd edition, Mathematical Surveys and Monographs, Vol. 120. American Mathematical Society.
[Sib]	Sibuya, Y. 1975. *Global Theory of a Second Order Linear Ordinary Differential Equation with a Polynomial Coefficient*. North-Holland.
[Sima]	Simader, C. G. 1978. Essential self-adjointness of Schrödinger operators bounded from below. *Math. Z.*, **159**, 47–50.

[Sj1]   Sjöstrand, J. 2003. Pseudospectrum for differential operators. Séminaire à l'Ecole Polytechnique, Exp. No. XVI, Séminaire Equations aux Dérivées Partielles, Ecole Polytechnique, Palaiseau.

[Sj2]   Sjöstrand, J. 2009. Spectral properties for non self-adjoint differential operators. In Proceedings of *Colloque sur les Équations aux Dérivées Partielles*. Évian.

[Sj3]   Sjöstrand, J. 2010. Resolvent estimates for non-selfadjoint operators via semigroups. In Laptev, A. (ed.), *Around the Research of Vladimir Maz'ya III*, International Mathematical Series, Vol. 13, pp. 359–384. Springer/Tamara Rozhkovskaya.

[SjZw] Sjöstrand, J. and Zworski, M. 2007. Elementary linear algebra for advanced spectral problems. *Ann. Inst. Fourier*, **57**(7), 2095–2141.

[St]   Staffans, O. 2005. *Well-Posed Linear Systems*. Cambridge University Press.

[Tr1]   Trefethen, L. N. 1997. Pseudospectra of linear operators. *SIAM Rev.*, **39**, 383–400.

[Tr2]   Trefethen, L. N. 2000. *Spectral Methods in MATLAB*. SIAM.

[TrEm] Trefethen, L. N. and Embree, M. 2005. *Spectra and Pseudospectra: The Behavior of Nonnormal Matrices and Operators*. Princeton University Press.

[Vi]   Villani, C. 2009. *Hypocoercivity*, Memoirs of the AMS, Vol. 202, No. 950. American Mathematical Society.

[Yo]   Yosida, K. 1980. *Functional Analysis*, Grundlehren der mathematischen Wissenschaften, Vol. 123. Springer.

[Zu]   Zuily, C. 2000. *Eléments de Distributions et d'Équations aux Dérivées Partielles*, Collection Sciences Sup. Editions Dunod.

[Zw]   Zworski, M. 2001. A remark on a paper by E. B. Davies. *Proc. Am. Math. Soc.*, **129**, 2955–2957.

[Zw2]   Zworski, M. 2012. *Semiclassical Analysis*, Graduate Studies in Mathematics, Vol. 138. American Mathematical Society.

Index

Printed in the United States
By Bookmasters